Epidemiology and Geography

Epidemiology and Geography

Principles, Methods and Tools of Spatial Analysis

Marc Souris

iSTE

WILEY

First published 2019 in Great Britain and the United States by ISTE Ltd and John Wiley & Sons, Inc.

ISTE Ltd
27-37 St George's Road
London SW19 4EU
UK

www.iste.co.uk

John Wiley & Sons, Inc.
111 River Street
Hoboken, NJ 07030
USA

www.wiley.com

Library of Congress Control Number: 2018965282

British Library Cataloguing-in-Publication Data
A CIP record for this book is available from the British Library
ISBN 978-1-78630-360-8

MIX
Paper from responsible sources
FSC® C013604
FSC www.fsc.org

Contents

Foreword

This book is the result of a long series of scientific works that the author has conducted for over 30 years. With an initial background in mathematics and computer science, Marc Souris is one of the few researchers who have focused their research efforts on the methodological development applied to spatial data, which he realized through many research programs involving various disciplines (geology, geography, epidemiology, etc.). Due to his quite unique positioning, his capacity to go beyond the frontiers of his initial academic training and his ability to clearly and objectively present principles that may seem complicated at first glance, this work has a particularly remarkable and unique character.

This book offers a very rich state of the art of the concepts, methods and tools of spatial analysis currently used in epidemiology and in certain works related to health geography, and that the author has intelligently organized in coherent chapters. This type of book is all the more valuable as overviews covering such a wide range of methods are rare. The author devotes particular attention to describing the formalism, the terminology and the scientific approach to be adopted by anyone willing to apply spatial analysis in the health field. The author warns, and with good reason, on the numerous pitfalls (confusion factors, ecological error, layout of spatial substratum, edge effect, etc.) and limits (oversimplification of reality, inadequacy between the level of analysis and the spatial scale of the processes, data reliability, uncertainty in localization, etc.) that have to be dealt with and for which solutions are proposed. The author uses compelling examples, particularly in relation to vector-borne infectious diseases, without, however, omitting other categories of diseases (notably chronic or degenerative diseases, such as long-term disorders or diabetes), although the latter are less often mentioned in this book. The examples refer to study sites predominantly located in southern countries (Latin America, Southeast Asia and Africa).

A further very significant aspect is that the methods are presented in a highly didactic manner, by means of simply formulated questions to which they allow an answer. Whenever possible, several software solutions are suggested in order to implement the advocated methods. Furthermore, it is worth noting that Marc Souris has himself optimized a number of methods presented in this book and has also developed new ones, among which is an operational implementation using the SavGIS free software.

Finally, this book features a balanced integration of theoretical and methodological issues, practical examples and elements related to the software to be used. There are summaries at the end of each chapter, numerous illustrations (maps and graphic representations), many boxed texts, a glossary, a rich bibliography and two detailed practical cases, all of these presented in a very accessible style, which facilitates the reading experience. Although it primarily addresses students enrolled in Master and PhD programs, researchers, research analysts or managers working in the healthcare sector, it is also a further reaching resource that can prove valuable for anyone willing to acquire knowledge on spatial analysis methods, regardless of the field of application.

Florent DEMORAES
Lecturer and Researcher –University of Rennes 2,
Deputy Director of UMR 6590 ESO
(Spaces and Societies – CNRS)

Preface

"I lie only to tell the truth"

Chinese proverb

This book gives an overview of the objectives, principles and methods of spatial analysis and of geographic information systems used in the healthcare sector, and particularly in the study of infectious diseases and of health–environment relations. It is designed as a practical introduction to spatial and space-time analysis for epidemiology and health geography. Its objective is to offer a detailed description of the objectives, concepts, and most of the methods and techniques available in this field, with a didactic approach illustrated by concrete examples. It is aimed at students and public health professionals, epidemiologists, public health inspectors, health geographers and experts in (human or animal) health–environment relations, who are interested in a comprehensive overview of the subject that does not require in-depth mathematics or statistics knowledge. Finally, the book also aims to be a tool that can be used by all of those interested in an introduction to the general methods of spatial analysis.

Spatial analysis includes any technique that studies objects and their attributes using topological or geometric properties, generally in a two- or three-dimensional metric space. This is a very general definition that applies to many domains. Spatial analysis is not a recent discipline; it has been used for many years in biology, botany, epidemiology, image processing, network analysis, electronic design, chemistry, cosmology, climatology, hydrology, economics, etc. Obviously, it has been used in geography, where spatial analysis is defined as "formalized analysis of the configuration and properties of the geographic space, as it is produced and experienced by human societies" [PUM 97].

In epidemiology (study of the factors influencing a population's health and diseases) and in health geography (geographic analysis of the health system and of the spatial distribution of diseases)[1], the term spatial analysis will be used to describe the analysis techniques applied to the "objects" described or used in epidemiology or geography, since they are localized in space and the analysis uses this localization: individuals, vectors, reservoirs, populations, territories, natural, urban or rural environment, etc. Spatial analysis uses topological or geometric relations of the individuals with their environment and among them. It is not concerned with what happens "inside" the sick person (in the organ, cell, or in terms of biology of the pathogen agent). For example, this book does not cover medical imaging and the techniques associated with image processing, although some of these techniques are sometimes very close to those described here.

Spatial distribution of health phenomena is rarely random: a health phenomenon often involves risk factors related to geographic factors, mesological factors and spatial relations among individuals. The use of localization is therefore essential in the analysis and comprehension of a health phenomenon and of its mechanisms. Spatial analysis facilitates the identification and comprehension of the mechanisms and processes that underlie the health phenomenon, by considering the spatial relations and interactions between the actors of the disease perceived as a complex system.

In epidemiology, spatial analysis also provides the elements that contribute to the consolidation of "traditional" epidemiology and feed the research and parameterization of models. It also enhances the analytical approach in health geography, whose methodological body also integrates a whole set of qualitative approaches. *Descriptive* spatial analysis includes cartographic analysis, search of geometric and space-time characteristics, analysis of the space variability of a value, cluster detection, spatial scale analysis, environmental correlation analysis, etc. *Explanatory* spatial analysis is essentially statistical, with the search of statistical models including spatial relations between individuals. Modeling of spatial processes is only briefly touched upon, this subject being beyond the scope of this book.

In health studies, spatial analysis is not only used for studies conducted in epidemiology or in geography. It also plays a role in public policies, with the development of new applications in public health: early warning systems, crisis management systems, risk analysis and prevention systems, preparation of vaccination campaigns, surveys and polls.

1 Precise definitions are provided in Chapter 1.

This book aims to present the general concepts that underlie spatial analysis and to explain and clarify the principles used in methods of analysis. Practical use of these methods is also highlighted: many concrete examples based on real data are provided throughout this book. These examples cover situations that are often encountered in practice.

In recent years, spatial analysis has been increasingly used in the health sector due to the development of geomatics and geographic information systems (GIS). In health, as well as in other fields of application, spatial analysis has benefited from the spread of GIS use, the development of their technical functionalities and the growing availability of geographic data, despite their often inadequate quality.

It is difficult, if not impossible, to manage, transform, handle, analyze and represent spatial information without using GIS.

Finally, I would like to thank all of those who have contributed to this book. Firstly, Jean-Paul Gonzalez, physician and virologist, who offered unequalled inspiration, motivation and management with unrelenting enthusiasm; Florent Demoraes, geographer, who contributed to the reinforcement, consolidation and completion of these reflections; Bernard Lortic, engineer, whose highly demanding approach was unparalleled; all the colleagues, students, PhD students and interns who have directly or indirectly contributed to the improvement of this book, and in particular Nitin Tripati, José Tupiza, Somsakun Maneerat, Julie Vallée, Jothiganesh Sundaram and Tania Serrano. I am taking this opportunity to express my sincere gratitude to all of them.

Marc SOURIS
December 2018

Introduction

Software and Databases

The reader will find throughout the text information on how to apply the methods presented in this book using several pieces of software that have been selected for this purpose. Several databases or file sets that can be downloaded and used for the replication of the examples mentioned in this book are also presented.

An appendix presenting the principles and diverse functionalities of geographic information systems (GIS) is available for the reader to download at www.iste.co.uk/souris/epidemiology.zip.

I.1. Software

Several software programs which can be used to apply the methods presented in this book have been selected: general geographic information systems (QGIS, ArcGIS, SavGIS) or more specific software programs (R, GeoDA, SaTScan™, GWR4). Alongside descriptions of methods of spatial analysis, procedures to be used and links to find information on these preocedures, whenever available, will be briefly presented for each software program, without further details. If needed, the reader can refer to the software user manuals.

I.1.1. *QGIS*

QGIS

Quantum GIS (QGIS) is a free and open-source geographic information system. It operates under Linux, Unix, Mac OS X, Windows and Android, and supports numerous formats (vector and matrix) of data and databases. QGIS offers a continuously increasing number of functionalities provided by the basic functions and plugins. Detailed information, documentation, downloads and tutorials are available at http://www.qgis.org.

I.1.2. *ArcGIS*

The ArcGIS geographic information system is a commercial product from the Environmental Systems Research Institute (ESRI). This software is quite comprehensive, consisting of a large number of functionalities. The system's infrastructure allows us to share maps and geographic information among an enterprise, a community or on the Web. Further information on the ArcGIS software can be accessed at https://www.arcgis.com.

I.1.3. *SavGIS*

SavGIS is a free geographic information system running under Windows. This complete and powerful software is the result of research and is constantly evolving, providing innovative solutions for processing of localized information, with many developments related to spatial analysis and modeling for epidemiology. Besides being freely accessible, it has many advantages: rigorous data management, data sharing, powerful analysis, advanced functions for spatial analysis and statistical analysis functions. Further information can be found at http://www.savgis.org.

I.1.4. *R*

R (free and open-source software) is a programming language for statistical analysis of data, and also an environment for data analysis and graphic visualization. Scientists and researchers have created a large number of specialized procedures for a wide variety of applications that are directly integrated in R. R-GIS.net is a website that aims to discuss spatial data manipulation and analysis in R. Several packages are available for procedures related to spatialized data: sp, spdep, etc. Further information can be found at http://r-gis.net and framabook.org/r-et-espace.

I.1.5. *GeoDA*

GeoDa is a free and open-source software for spatial analysis, developed since 2003 by the State University of Arizona (USA). GeoDa is a software tool focused on spatial analysis and spatial models. The program provides a user-friendly graphical interface for the exploratory spatial data analysis (ESDA) methods, such as spatial autocorrelation statistics for aggregated data and basic spatial regression analysis for punctual and zonal data. Further information can be found at http://geodacenter.github.io.

I.1.6. *SaTScan™*

SaTScan™ is a free software that analyzes spatial data by means of spatial, temporal or space-time statistics. The main objective of SaTScan™ is the detection of aggregates and the implementation of early warning or early detection of disease systems. The software can also be used for similar problems in other scientific fields. Further information can be found at https://www.satscan.org.

I.1.7. *GWR4*

Geographically weighted regression (GWR) is a spatial analysis technique that takes into account variables exhibiting autocorrelation and local variations. This regression technique allows the modeling of local relations between predictive variables and the variable to be explained. Several software programs allow the execution of geographically weighted regressions (ArcGIS, SpaceStat, SAM, spgwr, gwrr packages or GWmodel of R), but GWR4 is an autonomous

Windows application. Further information on GWR4 can be found at http://gwr.maynoothuniversity.ie/gwr4-software/.

I.1.8. *Gama*

GAMA is a free agent-based modeling and simulation platform developed since 2007 (http://gama-plaform.org), and more specifically dedicated to the simulation of spatialized phenomena. GAMA offers a certain number of advanced functionalities: advanced management of geographic data; a set of structures and controls facilitating the definition of multilevel models; automated tools that support the exploration of models allowing the definition of experience plans and their execution on high performance calculation resources (cluster, grid); a plug-in system that allows the extension of GAML language for specific needs; and bridges and possibilities of coupling with other tools used in the field of modeling of complex systems.

I.2. Data for the examples

The methods presented in this book are illustrated with examples drawn from real situations and databases. The data related to these examples are available as EXCEL files for non-localized data, Shapefile format for geolocalized data, or complete geographic databases directly exploitable with the SavGIS software. These files, as well as the SavGIS databases, can be downloaded from www.savgis.org.

1

Methodological Context

This introductory chapter presents the methodological context of spatial analysis applied to epidemiology and health geography. It introduces the systemic approach to health research, the notion of risk in this context, and the various areas of research that use the methods and tools presented in this book.

1.1. A systemic approach to health

A **health phenomenon** – the set of changes in the physiological or sanitary status of the individuals in a population linked to a pathology or a pathology-related characteristic – is the result of processes that are always determined by numerous parameters. Some parameters are affected by the individuals' personal characteristics (general characteristics such as age or sex, biological characteristics, genetic characteristics, etc.), and were for a long time the only ones used by biology and medicine[1]. However, a health phenomenon is also determined by factors linked to behaviors and interactions: mainly relationships between individuals (contacts, spatial proximity relationships, behavioral relationships) or relationships between individuals and their environment (natural, social, economic, etc.). The general objective of epidemiology is to understand and model these processes.

When studying or analyzing the characteristics of populations, it is difficult to understand behaviors and interactions, and equally difficult, if not impossible, to describe their entire complexity at the individual level. Some of the individuals' characteristics and their interrelations (for example, movements) are difficult to

1 The mesological approach associating man with his environment in the widest possible sense only emerged in the 19th Century as a continuation of the Lamarckian theory of interactions between biology and environment.

describe at the individual level: they are generally determined by probabilities, using statistical analysis of populations. Several levels of aggregation of individuals in a population are possible for their definition and description. These levels correspond to what is commonly known as a description scale, level or spatial granularity of data, concepts which simplify the empirical reality in a description model. The environment itself can be described at several scales, depending on how reality is modeled. Finally, individual characteristics can themselves be directly influenced by environment or behaviors.

The global approach to health issues therefore requires a systemic perspective, in which the sole medical aspect (biological and individual), although essential, is not by itself conducive to explaining the phenomenon or mastering the impact on the individual or on the society. According to the **systemic approach**, a health phenomenon is a complex system, involving various groups of "agents" that act and interact depending on their characteristics and environments, according to **processes** which we will aim to decode from observed situations, and then model. The various groups of agents consist of:

– **Individuals** (human or animal, potentially susceptible to being individually affected by the pathology or by the phenomenon, and to changing their health or physiological conditions);

– **Pathogens** (virus, bacterium, parasite, fungus, prion, etc.) in case of infectious diseases;

– **Toxic substances** or **pollutants** (asbestos, metals, radioactive products, chemical products, pesticides, etc.) that can cause certain non-infectious diseases;

– Possibly, **vectors** (animal that transmits the pathogen to the host, such as mosquito, tick, rodent, bird, carnivore, etc.);

– Possibly, **reservoirs** (animal that preserves and spreads the pathogen in the environment, while not necessarily being affected, such as civet, bat, bird, etc.).

In the case of infectious diseases, individuals (human or animal) are often called **hosts** or **potential hosts** when they are susceptible of being infected. Most of the pathogens are **mobile**, and are carried by a host, a vector, or a natural element (air, water), or by mechanical transportation means (airplane, ship, truck, etc.). Many pathogens are also present in soils, and can therefore be considered **immobile**, with the exception of sediments carried by a water stream.

The processes and mechanisms, which we are looking to model, and can enable the understanding of the health phenomenon as a whole, are considered to be global mechanisms, identical throughout the studied territory. Many **environmental factors** are involved in these processes and directly influence, when exposed to them, the characteristics of the various agents, their behaviors and their relationships as individuals or as groups of individuals. A spatial distribution of the phenomenon is the result of all of these processes.

EXAMPLE.– Temperature and rainfall influence the development of mosquitos, and therefore the transmission of a mosquito-dependent disease. Many viruses are sensitive to UV radiation and are rapidly damaged by a sunny environment.

The **health system** (care and prevention for humans and for livestock) is also one of the "environmental" factors influencing the characteristics of a disease.

Diseases which involve one vector (sometimes two) are called **vector-borne diseases**. They are obviously strongly dependent on the behavior of the vector, which is itself influenced by the environment. Many diseases do not involve pathogens (**non-infectious diseases**, such as diabetes, obesity, some cancers, growth abnormalities, etc.), but their study is not any less simpler, since it has been observed that individuals' behavior and environmental factors (in broad terms) can also have a significant influence on non-infectious diseases.

The systemic approach considers a **health phenomenon** as a complex system, consisting of various groups of "agents" that act and interact according to their characteristics and to their environments: hosts, pathogens, reservoirs and vectors (Figure 1.1). The health phenomenon can affect the state of a "host" and cause it to change from "healthy" to "sick" status. Complexity in studying a health phenomenon essentially arises from the dynamic aspect of the system and the interdependency of its components. Nonlinear interactions among elements may generate unexpected behaviors at a global level [MAN 16].

Box 1.1. *Systemic approach*

The agents and environmental variables used to describe this system and that have an influence on the health phenomenon (increasing the disease probability) are called **risk factors**. These risk factors, and in particular the environmental ones, can be highly variable in space and time. Events of low or very low probability must sometimes be taken into account, which may potentially result in high instability of the overall system, and make a purely deterministic approach difficult, if not impossible, especially at the individual level. If process analysis and modeling (*why, how*) is nevertheless achieved, this random instability rarely makes it possible to fully predict a phenomenon (*who, where, when*). In these cases, we are able to calculate the probabilities for only some of the health phenomenon's characteristics, and most often for groups of individuals rather than for individuals: the model allowing process simulation should involve many stochastic elements.

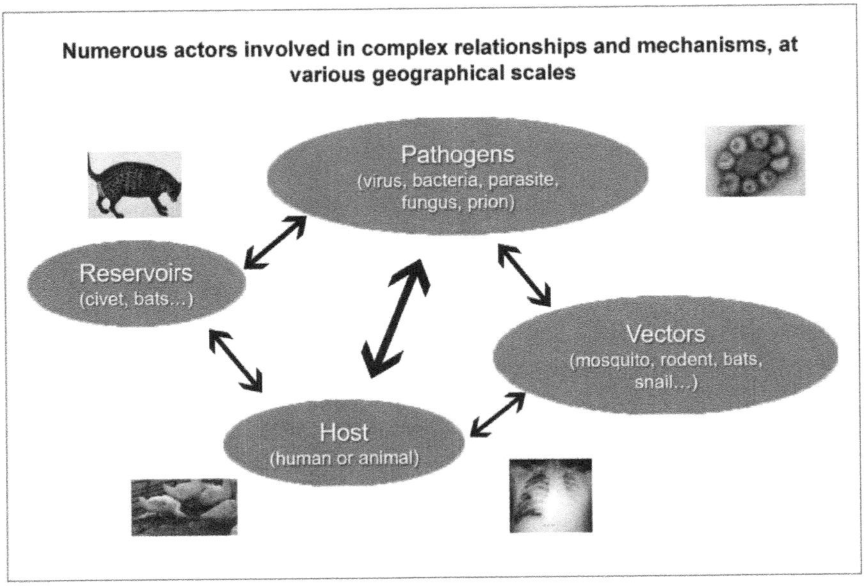

Figure 1.1. *An infectious disease is a particularly complex system: numerous actors involved in complex mechanisms, at several scales, all interrelated, and in relation to their environments*

This systemic and multi-factorial perspective has led health research to become largely **multidisciplinary**. While medical research usually focuses on the medical and biological aspect of a health phenomenon, at the level of the individual, treated as a patient, health research now involves many disciplines, for which the individual is not necessarily a patient, nor the main focus of study. The health system also plays a specific role: it is simultaneously the central factor influencing a health phenomenon (since it seeks to manage and reduce it), and at the same time it is key for collecting epidemiological information used to evaluate and analyze this phenomenon (at the population level) and to measure its own impact. It should be kept in mind that in epidemiology, data reflect the effects of the disease (measured by the health system), and not the disease itself.

Health research involves many disciplines, including, in particular:

– medicine for the study of pathologies, patients, care and treatments;

– biology and virology for the study of pathogens;

– epidemiology for the study of etiology and risk factors, with a population-based statistical approach;

– entomology, biology, ecology, zoology for the study of vectors and reservoirs;

– ecology and geography for the study of the environment;

– social sciences (geography, anthropology, sociology, economy) and geomatics for the analysis of the health system, resource analysis and optimization, characterization of vulnerabilities and the study of their mechanisms;

– mathematics, statistics, information sciences for phenomenon characterization, process modeling, development of monitoring and early warning systems.

Box 1.2. *Health research involves many disciplines*

Spatial analysis is used in the systemic study of a health phenomenon as most of the actors (agents, environmental factors) are localized in space and in time, and many relationships are proximity-based. The use of spatial analysis for an observed situation contributes to determining and characterizing the processes and factors that generated it. As will be seen throughout this book, **geomatics** (a disciplinary field based on data processing concepts, tools and methods that allow the acquisition, management, representation and analysis of localized data) methods and tools are essential in the practical implementation of spatial analysis. Many references to the software concerned are provided in this book. Geomatics allows, in particular, the management of the influence of geographic levels and context of this complex system, where elements can often be described at various geographic scales.

Since localization measurements have become quite simple technically with positioning systems (like GPS or Galileo systems), geomatics has contributed to most of these disciplines, facilitating the development of many scientific or business applications.

1.2. Risk and public health

Once this systemic framework is defined, a "**risk**" perspective can be adopted, in which various elements of the system (agents and environments and their variables) are classified according to their estimated influence on the probability of the health phenomenon at the individual level – the **risk**, considered as the probability of disease or of disease effect [OMS 02]. This pragmatic approach makes it possible to structure the scientific method analyzing the health phenomenon, to rationalize and enrich the description of agents and their environments, in particular through the notion of vulnerability. Above all, it makes it possible to rationalize prevention and risk reduction actions by adopting a "public health" approach. It enables the focus on results which can be directly used in public health policies without requiring the analysis of all the processes involved in the studied phenomenon. Moreover, most epidemiological studies aim to investigate risk factors rather than decipher and model all the processes.

In this classification, we distinguish what is threat-related and what is under vulnerability (that is, the capacity to be more or less affected by a threat):

– The **presence of a threat** (or "**hazard**"), which can be a pathogen, a vector, a reservoir, and also pollutants, toxic substances, noise, industrial presence, etc. These elements are considered necessary – but never sufficient – for the development of the health phenomenon. They are often known only in terms of probabilities, which are sometimes very low, and they are potentially subjected to significant random variability in time and space. Actually, temporal or spatial situations with a very low probability of occurrence are often encountered, which confirms the interest of spatial and space-time analysis: very often, the objective of studies is to evaluate the spatial and temporal differences of such a probability, even though very low, in an attempt to measure its significance. Sometimes only a characteristic required by the pathogen or vector presence is used (for example, water presence, a minimum temperature or a type of vegetation).

– The **susceptibility** of the individual (essentially due to individual, genetic or biological characteristics, such as immune status or age, and strongly related to the pathology). It is an individual, and often provided by a probability. Susceptibility is a form of "**inevitable**" **vulnerability**, on which it is often difficult, if not impossible, to act (aside from vaccination, the ultimate weapon against infectious diseases, which allows an individual to be rendered non-vulnerable and makes it possible to reduce the susceptibility of a population).

– "**Passive**" **vulnerability** of the individual, which is not directly dependent on pathology, which is neither necessary, nor sufficient, and which influences the individual exposure to hazard or its protection against pathology. Protection consists of prophylaxis, access to healthcare system and response to treatment. It is independent of the real presence of the hazard: one can be vulnerable without being exposed to the threat. The definition of vulnerability often unfolds on several levels (individual, contextual). It is very often **spatialized** because it is related to geographical segregation or spatial concentration phenomena. This field is essentially studied by geography. This is the main target of public policies to reduce risk.

– "**Active**" **vulnerability**, which includes all the factors that are susceptible to increasing the direct individual **exposure** to the hazard. Active vulnerability consists of the so-called "risk behaviors" increasing the individual's probability of encountering the hazard, by exposing it to an environment where the hazard is present (for example, movements and contacts, professional activities, risk practices). The identification of active vulnerabilities in populations allows the optimization of risk reduction policies, by targeting the groups concerned.

Risk is the **conjunction** of the hazard presence with these various vulnerabilities.

> **Vulnerability** is the quality of that or who is susceptible of being affected by a threat. The concept can be applied to a person, group, object and space, and it indicates its ability to prevent, face or resist a threat.
>
> The term vulnerability is used in many disciplines: economy, health, nutrition, law, sociology, environmental sciences, etc. It is widely used by major international bodies.
>
> When sector based, vulnerability is expressed by "variables" (called **vulnerability factors**) determined in relation to a specific threat. These vulnerability factors can be individual, contextual, structural, etc. The notion becomes universal when the threat is no longer specific, but it is itself a set of sector-based threats. Statistical indicators have been proposed in order to quantify this notion in relation to various vulnerability factors and facilitate its use.

Box 1.3. *The notion of vulnerability*

By introducing the concept of risk in the systemic approach of a health phenomenon, there is a shift from a **"medical" perspective**, focusing on the treatment of disease effects, towards a **"public health" perspective**, focusing on risk reduction and improved well-being. According to the latter, the study of exposures and vulnerabilities appears in its full significance, in the context of medical research on treatments and of a healthcare system study. This approach makes it possible to rationalize prevention and risk reduction policies. It distinguishes between natural biological threat, often subject to high random variability in time and space (emergence or presence of the hazard, which is quite often difficult, if not impossible, to control), and vulnerability, which is generally far more stable at the level of populations (susceptibilities, exposures, behaviors and vulnerabilities), and allows for better public health policies. It also facilitates a rational management to crisis on the one hand by creating preventive interventions targeting the most vulnerable elements of the system (**issues and challenges**), and on the other hand by optimizing risk reduction policies in emergency situations (by threat elimination – elimination of vectors, quarantines, slaughters, etc. or by susceptibility and exposure reduction – vaccinations, hygiene, protection).

EXAMPLE.– The biosecurity of livestock farms is a major issue in the management of epizootic diseases; maintaining *herd immunity* above a threshold is essential to prevent the development of an epidemic.

In any case, these interventions must be adapted to the social context, to ensure significant impact on the risk behaviors and exposures they induce, hence the growing role of anthropology in the field of health. Risk reduction measures that are not socially accepted may have disastrous consequences, as can often be noted during vaccination campaigns.

– Reducing individual susceptibility (immunization, vaccination, prophylaxis).

– Reducing individual exposure to the pathogen (vector control, reduction of pathogen- or vector-favorable conditions, quarantine and exclusion zone).

– Eliminating the pathogen, either directly (slaughter, disinfection, hygiene) or indirectly (by suppression of transmission).

– Reducing individual vulnerability (social and economic conditions, behavior, prevention and better access to care system).

– Reducing exposure in emergency situations: implementation of data collection and sharing systems, implementation of early warning and crises management systems, implementation of treatment and observation centers.

Box 1.4. *How to minimize a risk?*

Treatment availability or complete knowledge on the pathogen is not sufficient to eliminate a disease. Even if an element contributes to a significant risk reduction (for example, a highly effective vaccine, such as the one for yellow fever), other system agents should be considered to ensure its effectiveness in a risk situation,. When the system involves an animal reservoir (as it is the case for yellow fever, influenza, rabies, dengue fever, etc.), the pathogen cannot be eradicated. Many human pathogenic bacteria are present in the environment and will never be eliminated (for example, for diseases such as tetanus, cholera, leptospirosis, anthrax, etc.). To date, only the smallpox virus has been eradicated due to host susceptibility reduction by vaccination campaigns, and this has led to suppressing transmission and therefore the pathogen, as the latter had no wild reservoir. Eradication of the measles virus thanks to vaccination campaigns and the subsequent interruption of transmission were part of the WHO objectives for the decade 2010–2020; however, their achievement is unlikely, as evidenced by numerous instances of resurgences of the disease, mainly due to insufficient vaccine coverage.

Data-based **risk estimation** is at the core of epidemiology and statistical modeling in epidemiology. Using mainly a statistical approach, epidemiology seeks to determine the variables of agent and environment that can influence the health phenomenon, and to use these variables for modeling the global health phenomenon.

The term **risk factor** refers to the agent and environment variables that can influence the health phenomenon (threat or vector presence; individual susceptibility; threat exposure; vulnerability). Evidencing the risk factors of a disease based on case-related data is one of the main objectives of epidemiology.

Box 1.5. *Risk factor*

This "risk" approach and the resulting expression in terms of hazards and agent or environment vulnerability involved in the systemic description of a health phenomenon can be found in most epidemiology studies. It contributes to defining a conceptual framework and guiding the investigation process. Most of the examples presented in this book illustrate this approach.

Spatial distribution of the disease effect (studied on the basis of observed data) provides significant information on the processes, both on the influence of certain variables, on the research of environmental risk factors and on the influence of relationships between agents. Explaining spatial differences by means of risk factors is at the core of spatial analysis of risk, especially when agents or environments are considered immobile. This is particularly the case when data are aggregated by geographic units, which will be studied in detail in Chapters 4 (Cartographic Representations) and 6 (Spatial Analysis of Risk).

1.3. Epidemiology

The main objective of **epidemiology** is to describe and measure the characteristics of a pathology or of a state of health in a population, to estimate the risk, to determine the causes of this pathology or state of health by a statistical approach, based on the data observed in this population [BOU 93]. Epidemiology differs from a purely biological and medical approach that seeks to explain an individual's state of health by providing a description of the biological mechanisms of the disease at the individual level. The epidemiology approach to the study of diseases is quite recent, and it relates to the rapid expansion of the theory of probabilities and statistics at the beginning of the 20th Century. As in all the fields of application of probabilities and statistics, many traps and pitfalls must be avoided in order not to reach false conclusions: context effects, counts, dependencies between variables, probabilities, prediction, rare events, confusion, bias, dependencies between events, etc.

Epidemiology is essentially a quantitative and statistical (descriptive or inferential) approach based on qualitative or quantitative data collected on individuals or groups of individuals in the population affected by the studied phenomenon: this collected data allows us to calculate prevalences, incidences, pattern of the distribution of a characteristic, make group comparison, analyze the relationships between disease and exposure to a factor, analyze differences between patients and non-patients, etc. The main objective of epidemiology is to uncover the relationships between disease and risk factors, and to confirm the presence of mechanisms through which risk factors affect the disease (using statistics, mathematics and modeling). Nevertheless, epidemiology does not seek to directly explain how these relationships and their mechanisms occur. This is the scope of other disciplines, mainly biology, geography, sociology, anthropology, etc.

For epidemiology, the location of a phenomenon is not relevant in itself. The spatial characteristics of the processes must be statistically explained, based on data and using environmental variables and interactions between agents, or between agents and environment. Localization is used to connect agents or factors in order to answer a specific question with a statistical approach. Therefore, spatial analysis in epidemiology can provide other disciplines with elements which can enhance their understanding of a phenomenon.

EXAMPLE.– Even though no zone features an abnormally high incidence rate, the detection of a cluster of zones with high incidence (but not abnormal if considered individually) can uncover the presence of a phenomenon at another geographical scale of aggregation.

In a classical statistical approach, observed values are always considered as being part of a set of possible values (whose calculation relies on *a priori* hypothesis). Most analysis techniques used in epidemiology, either spatial or not, rely on this principle. Rather than describing an observed situation, the aim of epidemiology is to evaluate the probability of this situation occurring with respect to all possible situations. At the heart of the epidemiology statistical approach is to show that the observed situation is unlikely to have occurred by chance.

1.4. Health geography

Health geography is a synthesis approach, at the core of which is the territory (space that results from a multifactorial construction). The approach to public health issues (diseases and health system) often relies on the study of territories, and not the reverse. Health geography covers a wide field of study, which includes domains not covered by epidemiology.

Health geography covers the geography of diseases (analysis of the spatial and social distribution of diseases), the geography of the health system (localization of healthcare facilities, analysis of the spatial distribution of the healthcare system, spatial disparities in the health system, accessibilities, inequalities, flow studies, utilization of health services, hospital attractiveness models, etc.), the geography of populations and territories in relation to health (health assessment, vulnerabilities, planning, resource allocation, influence of health in the geographical construction of territories) and regional health planning (identification of needs and priority objectives, predictions, definition of health areas and health territories). It also encompasses the study of public health issues related to behaviors and infrastructures (accidentology).

EXAMPLE.– Whether it is a matter of incidence of diseases or mortality, child health, cancer risk, perception of health or access to the healthcare system, a systematic disadvantage can often be noted in the case of socially deprived groups. Social segregations, spatial segregations and health segregations are quite often interrelated. Evidencing spatial inequalities in health and detecting their determining factors are an important objective for health geography.

By definition, geography is particularly interested in the places and territories of phenomena. For geography, the place is meaningful, as a synthesis of a set of structured and unstructured mechanisms. Geography gives meaning to space: it attempts to explain the operating rules of the system, with both quantitative and qualitative tools. Furthermore, geography attempts to explain spatial characteristics through synthetic analysis.

Geographical and epidemiological approaches are therefore complementary. Their directions are often opposite, and they do not use the same methods: epidemiology uses an analytical approach to analyzing a complex problem by reducing it to several simpler problems, while geography tries to generate a synthesis of various aspects of a health phenomenon, and to define a system that facilitates the overall understanding and explanation of the phenomenon.

Although health geography and epidemiology differ in terms of approach, when it comes to using space, the two disciplines share many of the same tools, and in particular those of spatial analysis, statistics and geographic information systems.

1.5. Spatial analysis for epidemiology and health geography

The geographical and temporal localization of the different "agents" provides significant information for the study of a health phenomenon: localization is directly involved in the relationships between agents in the system, in the relationships between pathogens and susceptible hosts, in the direct relationships between individuals, and in the exposure to geographical or environmental risk factors.

As has been already mentioned, the processes to be explained will be considered global mechanisms, identical throughout the studied territory. The process result depends on all the factors and events (deterministic and random) that are involved in these mechanisms. The analysis of the spatial distribution of the observed phenomenon aims to determine or characterize the processes and their factors, based on the spatial characteristics of this distribution.

There are thus several reasons why considering and studying the spatial or space-time distribution in a health phenomenon is important. Taking into account localization makes it possible to improve the search of risk factors, generating information on the relationships between the health phenomenon and the environmental conditions in which the phenomenon occurs. In more general terms, spatial distributions make it possible to rapidly formulate hypotheses on the mechanisms and processes underlying the studied phenomenon, as soon as these mechanisms and processes involve spatial relationships (between agents or groups of agents, with environmental factors, at various geographical scales). Finally, by producing localized alerts, taking location into account makes it possible to control a contagious phenomenon by acting on the transmission, as long as this transmission remains limited to a small area. Spatial analyses can thus rapidly lead to deciphering the processes related to several key risk factors and allow concrete mitigation or prevention actions in public health.

EXAMPLE.– Spatial relationships between disease and environment are at the center of the study of risks of certain diseases related to pollution sources (such as some cancers or hormonal disturbances). Taking into account localization then becomes essential to understand the phenomenon.

The influence of spatial relationships between "agents" is especially important for infectious diseases, since by definition contagion-based transmission of a pathogen involves proximity or contact. Nevertheless, non-infectious diseases (obesity, diabetes, some cancers, etc.) may also have specific spatial distributions. These spatial distributions may be due to the exposure to environmental factors themselves having specific spatial distributions, such as clusters (for example, pollutions related to stationary punctual sources). Spatial clusters can also be due to clusters of behaviors, vulnerabilities or susceptibilities, the identification and analysis of which are essentially within the scope of geography and anthropology. Human health is thus closely linked to space organization and to human behaviors attached to it, which are themselves spatially structured. Conversely, human behaviors and space organization are quite often built and structured according to constraints related to health, hygiene, healthcare availability, nutrition, security, transportation, etc. Geographical analysis of vulnerability factors or of hazard presence allows us to highlight societal relationships resulting from the organization of space by society, and to think about political actions allowing their reduction.

Like statistics, spatial analysis can be either descriptive or explanatory. Spatial analysis provides a set of tools aimed at uncovering or highlighting spatial and temporal differentiations in the distribution of events, and at testing the hypotheses on the factors causing these events by using their localization in time and space. The main objective of spatial analysis in epidemiology is therefore to facilitate the identification of localized risk factors and their distribution of probability, thus providing the elements for characterizing and modeling processes. It allows the visualization, synthesis and analysis of positions and spatial relationships between events (continuity, clustering, attraction-repulsion, pattern, centrality, movement, diffusion process, etc.). Furthermore, it allows the analysis of the relationships between the spatial distribution of the values of an attribute and the environmental characteristics of the phenomenon (environmental correlations). It also allows us to consider spatial structures of risk factors in the explanatory statistical models. Tools have therefore been developed in order to:

– visualize spatial distributions, with description and visualization tools, allowing the visual analysis and synthesis of the observed situations;

– synthesize and analyze the positions and spatial relationships between events (continuity, clustering, attraction-repulsion, pattern, centrality, movement, diffusion process, etc.);

– analyze the relationships between the spatial distribution of the values of an attribute and the environmental characteristics of the phenomenon (environmental correlations);

– model the emergence, diffusion, epidemic extinction, with process modeling and simulation tools, in order to evaluate the various possible situations according to the hypotheses elaborated during the analysis.

Spatial analysis essentially serves to:

– provide information that contributes to explaining a phenomenon and identifying the corresponding risk factors;

– analyze the observed processes, and define their spatial characteristics and parameters allowing their modeling.

The main tools for spatial analysis are used for:

– mapping of the results of epidemiologic analyses by geographical unit after spatial aggregation: epidemiologic indices, absent or excessive risks, residues, etc.;

– spatial analysis of observed situations or of statistical residues after a regression: position, distribution, characteristics of distributions or of spatial relationships (global or local), characteristics of local density, pattern, centrality, etc.;

– statistical and spatial analysis for the identification of risk factors: using localization of individuals or groups of individuals for the study of risk factors;

– analysis of observed or simulated space-time processes;

– spatial or space-time modeling of the processes and spatial analysis of simulated situations.

Box 1.6. *Spatial analysis in epidemiology and health geography*

Spatial analysis for epidemiology and health geography consists of two large groups of methods: those analyzing the localization of events themselves, and those analyzing the geographic units in which the events have first been aggregated depending on their localization. The first group of methods makes it possible to take into account the spatial relationships among "agents", and also requires data on the individuals, which are often difficult or impossible to obtain. These will be presented in Chapter 5. The second group of methods does not allow taking into account the direct spatial relationships among "agents", but it replaces the individuals by groups (geographic units), and therefore allows the use of data on groups, without requiring specific data on the individuals. These methods will be presented in Chapter 6.

Among the studies conducted in public health, the following are particularly worth mentioning:

– Health data visualization and health atlases. Mapping allows us to represent what is happening in each place, but it is essentially used in health in order to understand or illustrate a particular trait in the global or local spatial distribution of the studied phenomenon, such as a gradient, a pattern, a spatial tendency or a cluster. It essentially uses data aggregated in geographic units, whose rates can be calculated. It is a natural tool which allows the illustration of a geographical or statistical analysis concerning the global or local spatial distribution of the phenomenon. Highlighting inequalities in health, differences in accessibility and the analysis of their determining factors constitute an important direction in health geography. A simple cartographic representation shows that quite often health conditions and medical practices are not randomly distributed over a territory. Health atlases thus allow the synthetic presentation of territorial disparities. The notes accompanying maps contain geographical analyses of the factors that induce these disparities: these notes are essential, as mapping cannot be used by itself, in the absence of analyses allowing interpretation. These geographical analyses use classical statistics and spatial analysis.

The National Health Service (NHS), the state health system in the United Kingdom, has implemented an "NHS atlas" which enables a visualization of variations in activities and healthcare costs in order to optimize care efficiency (http://www.rightcare.nhs.uk/). The variations observed indicate the need to focus on certain care services and to study the possibilities for "overuse" or "underuse" of certain interventions, and for valorization of less costly activities. This approach provides the public health managers and decision-makers with tools to maximize results and minimize inequalities in care.

In the United States of America, the "Dartmouth Atlas Project" uses MEDICARE data in order to evidence inequalities in care access and to analyze the use of health services (http://www.dartmouthatlas.org/).

In France, the regional health agencies (*agences régionales de santé* – ARS) in cooperation with the technical agency for information on hospital care (*agence technique de l'information sur l'hospitalisation* – ATIH) provide interactive maps for health professionals, with detailed figures on care supply and demand by region, department, and also by district and municipality. Nevertheless, these tools are not connected to other datasets. They integrate neither the care accessibility outside hospital activities, nor the social and economic environment of the territories, which is required for a better understanding of the health-determining factors.

Box 1.7. *Interactive health atlases*

– Analysis of the access and use of the health system, optimization of the healthcare system and geomarketing of health. Numerous parameters are involved in these analyses: care offer, potential accessibility, observed accessibility, resources in transportation, offer and demand characteristics, demography, socio-economic conditions, etc.

– Studies of environmental correlations. The objective is to study the relationship between a health indicator and an environmental exposure. The exposure is either known for each individual, and the study is statistical (for example, cohort study), or the individuals are aggregated into spatial units that become the objects under study. The results of these studies should not be used at the individual level in order to avoid the occurrence of an ecological error[2] due to intra-unit variability, not taking into account the confusion factors at the individual level or not taking into account the time factor.

– Local studies around a source point or along a network. These studies focus on uncovering or highlighting a phenomenon (characteristics, disease, mortality) around sites exhibiting an assumed risk (as pollution or industrial risk). The population living in the proximity of the source is considered exposed, and is compared with a supposedly not exposed population, using an epidemiologic approach.

– Studies for the detection of places assumed to generate a health phenomenon, or of those that gather its consequences. These studies (generally relying on the observation of significant differences between theoretical values and observed values for a large number of cases, prevalences, incidences) focus on the detection of a place (case/prevalence/incidence that differs significantly from a theoretical value), a concentration, a cluster, a central place, a specific phenomenon pattern, indicating the possible existence of a cause–effect relationship between the geographical characteristics (environmental, industrial, demographic, sanitary) and the observed phenomenon. Independently of the search for causality factors, real-time detection of these places by monitoring and early warning systems contributes to limiting the propagation of a phenomenon.

2 See Chapter 2.

Early warning systems are currently used in many areas, particularly in weather forecast, fires, road traffic and natural disasters (tsunami, volcanic eruptions, earthquakes, etc.). Some of them allow real-time collection of information. In the health field, the WHO has implemented various early warning systems (global public health information network – GPHIN, global outbreak alert and response system – GOARN and GROG). The Heat Health Warning System (HHWS) has been particularly effective in reducing the mortality caused by heat waves. In France, the SENTINELLES network (Inserm, Pierre and Marie Curie University, French public health) (websenti.u707.jussieu.fr/sentiweb) allows the follow-up of space-time evolution of several infectious diseases (influenza syndrome, chickenpox, acute diarrhea, etc.).

Early warning systems use geomatics and space modeling methods: collection of localized data in real or quasi-real time (dedicated systems, social networks, participative methods), spatial analysis and detection of space-time anomalies (cluster, hot spot), real-time consideration of environmental conditions (in particular temperature, rainfall), spatial and dynamic distribution of vectors, socio-demographic conditions and consideration of the vulnerability of exposed persons, consideration of the access to the healthcare system and its capacities, etc.

Box 1.8. *Early warning systems*

1.6. Geographic information systems

Spatial analysis is based on the exploitation of localized data. Access and management of localized data have been made easily accessible by **geographic information systems** (GIS).

The development of concepts, methods and techniques related to geographic information, covered by the term **geomatics** (sciences of geographic information, geographic information systems, remote sensing), has reinforced the development of spatial analysis for geography in general, and for epidemiology in particular. GIS requires structuring information according to a rigorous data model. Applied to geographical data, this structuring allows the representation of reality by a data model that also manages the geographic localization. GIS brings together a large amount of localized data, at various scales, with various validities and accuracies, and various forms of description (for example, geometric descriptions using polygons or descriptions per pixel). They free the user from the complex task of technical data management, allowing him, once the data has been arranged and formatted, to analyze them through a process in which localization is easily and permanently available. The development of georeferenced remote sensing has greatly facilitated full access to environment knowledge, at many spatial resolutions. Besides the management of geographic information, most GIS now offer numerous tools for graphic representation and spatial analysis.

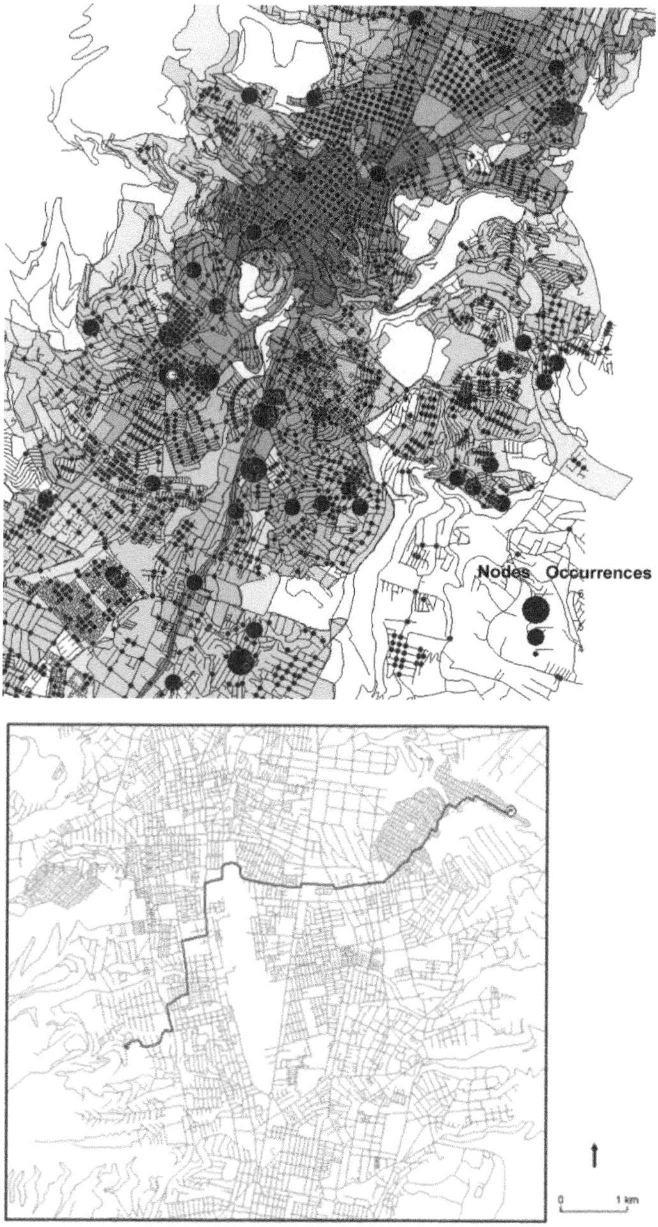

Figure 1.2. *Network analysis and calculation of shortest paths are examples of methods that are widely used in the analysis of health systems. For a color version of this figure, see www.iste.co.uk/souris/epidemiology.zip*

There are many data management methods and tools available in GIS, and they can be used for data analysis in health geography. Few geomatics techniques used in epidemiology or health geography are specific to this discipline, but some geomatics methods are much more widely used than others, while some are not used at all. The following presents a number of spatial analysis methods more specifically used in epidemiology and health geography, but the field of analysis of health geography is very wide (Figure 1.2). All the processing and spatial analysis methods available in GIS can be used, and knowledge on general GIS methods is essential.

There are many available software programs, either in the commercial offer (ArcGIS, MapInfo, etc.) or in the public domain (QGIS, GRASS, etc.). The GIS software used to illustrate this book (SavGIS) and complete examples can be downloaded from the SavGIS website (www.savgis.org). Guidelines for practical use of other software will be provided throughout the book.

The Appendix available online at: www.iste.co.uk/souris/epidemiology.zip presents GIS principles and their main functionalities. The reader is invited to refer to it at any time.

1.7. Book structure

Chapter 2 presents the general concepts of spatial analysis in epidemiology.

Chapter 3 presents various sources of localized data used in epidemiology.

Chapter 4 presents the methods and tools for the visualization of data used in the health field.

Chapter 5 offers a detailed presentation of the main methods used in the analysis of the spatial distribution of events, for epidemiology and health geography.

Chapter 6 presents the methods used for the spatial analysis of risk. These are classical statistical methods used in epidemiology with aggregated data, and in particular with environmental data resulting from the use of geographic information systems.

Chapter 7 presents the space-time analyses and the spatial modeling.

SUMMARY.–

– Health phenomena are understood from a systemic point of view, with multiple interacting agents that react to the environment, and according to their individual characteristics.

– Data-based risk assessment and prevention are essential objectives in public health. The study of spatial differences in risk is at the core of spatial analysis in epidemiology.

– Spatial relationships among agents and between agents and the environment may prove important in understanding a health phenomenon.

– The objective of spatial analysis in epidemiology is to highlight spatial differentiations in the distribution of events and to test hypotheses on factors involved in these distributions, in order to characterize and model the underlying processes.

– Geographic information systems structure and manage geographic data. They allow the implementation of spatial analysis.

2

Spatial Analysis of Health Phenomena: General Principles

As a preliminary step to a detailed presentation of spatial analysis techniques and tools used in epidemiology and health geography, this chapter aims to formalize various aspects: the needs (what should be determined), the observed situations (what is analyzed) and the methodology (how to use the methods and tools in order to meet the needs). A further objective of this chapter is to refresh the knowledge required for understanding these methods, particularly in epidemiology and statistics.

2.1. Spatial analysis in epidemiology and health geography

2.1.1. *Spatial distribution of a health phenomenon*

The overall **spatial distribution** of a health phenomenon (that is, the spatial distribution of the individuals "affected" by the phenomenon) results from many parameters, events and processes. To understand the spatial distribution of a health phenomenon, the initial distribution of "agents" should be first taken into account, and in particular the spatial distribution of "hosts" (individuals), and their susceptibility. A further step is to consider the characteristics of individuals or groups of individuals, the relationships between individuals and their environment (in a broad sense) as well as the direct interactions between individuals, particularly in a contagious phenomenon (for example, proximity or encounter), all these being involved in the processes that cause the events.

In general, the objective of statistical and spatial analysis of a health phenomenon is to identify and characterize the various events and processes involved in the studied phenomenon, and transform the spatial distribution of "hosts" (and other agents) into a spatial distribution of "patients". These events and processes are the following:

– **Processes of exposure** to risk factors, that is, factors that increase (or reduce, in the case of protection) the probability of the phenomenon occurring at the individual level. Risk factors can be individual (genetic, behavioral) or environmental. Environmental risk factors can be mobile or immobile, and often have a specific (non-random) spatial or temporal distribution, corresponding to their geography. Mobile factors imply a spatial distribution that varies in time while nevertheless preserving its specific spatial characteristics (geometric or topological).

– **Contagion processes** (or conversely, **inhibition processes**) that rely essentially on proximity or encounter (between individuals, between an individual and a vector, between an individual and a reservoir, between a pathogen and a vector or a reservoir, etc.). The factors that determine these processes are not individual, but are solely linked to the relationships, particularly spatial ones, between various actors (active or passive).

The spatial distribution of a health phenomenon thus results from several components, stemming from the spatial distribution of various agents:

– Spatial distribution of elements (agents, environments) involving a factor that is **causally related** to the phenomenon (for example, rain and proliferation of mosquitoes). These risk factors often feature spatial distributions that are not random, especially when they describe the environment (few natural phenomena have a random spatial distribution). The spatial distribution of a risk factor can be non-random in terms of its position and structure (for immobile factors) or only its structure (for mobile but structured factors: position can randomly move in time, while a structure persists). If an explanatory variable of the studied phenomenon has a non-random spatial distribution, this will influence the spatial distribution of the phenomenon, adding a non-random component. Conversely, through the analysis of the spatial distribution of a phenomenon, a specific environmental factor may be identified if its spatial distribution is not random and can be found in the observed distribution of the phenomenon.

– In contrast, if the studied phenomenon is correlated with a variable that has a random spatial distribution (or rather that cannot be differentiated from a random situation), its spatial distribution will be even more variable (random). Furthermore, if several non-correlated explanatory variables are linked to the studied phenomenon, these relationships will increase the variability of phenomenon's spatial distribution, which can rapidly become indistinguishable from a random distribution. The randomness of a spatial distribution is simultaneously caused by a purely random spatial component (similar to throwing rice grains over a tiled floor, and then counting the number of grains on each tile), and by the "mixture" of spatial distributions of all the factors that are causally linked to a phenomenon.

EXAMPLE.– The presence of bugs related to Chagas disease is partly linked to the presence of trees: on the one hand, if trees are not randomly distributed over the studied space, this spatial distribution will influence the spatial distribution of the disease. On the other hand, if trees are randomly distributed over the studied space, this random distribution of the risk factor increases the spatial variability of the health phenomenon.

The spatial distribution of a phenomenon is thus directly linked and results from the spatial distribution of factors that are causally linked to the phenomenon (risk factors). Each risk factor introduces variability in the resulting spatial distribution and increases the overall variability. It increases the uncertainty between the spatial distribution of another risk factor and that of the phenomenon. With localization (as with any multifactorial phenomenon), the maximum likelihood principle (what is observed is considered as most likely to occur) should therefore be used cautiously when applied to a single risk factor.

The purely random component of spatial distribution, "spatial noise", is the reference situation that should be achieved after having eliminated the influence of all known causal factors whose spatial distribution is not random (in terms of position and/or structure).

2.1.2. *Spatial analysis in epidemiology*

Spatial analysis of a health phenomenon mainly amounts to analyzing the spatial distribution of the phenomenon, in order to highlight and characterize, if possible, the underlying processes for explanatory or predictive purposes. This analysis comes down to characterizing the geometric attributes of the distribution (patterns, tendencies, clusters, homogeneity, continuity, etc.), characterizing the spatial variability of the phenomenon (and the spatial differences of this variability), highlighting the relationships between spatial distribution and estimated risk factors, and testing hypotheses on the factors involved in the processes. Spatial analysis complements traditional statistical analysis in epidemiology (traditional insofar as it does not directly consider the spatialization of objects).

Spatial analysis provides answers to many questions related to underlying processes such as:

Is the distribution of events or values random, clustered or uniform?

Is the phenomenon continuous in space?

Does the distribution of events or values involve clusters?

Is it possible to identify specific patterns and geometric structures?

Does the phenomenon have the same spatial characteristics throughout the studied domain?

Do the risk factors have the same influence throughout the studied domain?

Is it possible to identify interactions between events or between values?

Is there any correlation between the spatial distributions of events of two phenomena?

Box 2.1. *Spatial analysis*

Due to spatial and space–time analysis, space and time can be used in the formulation and testing of hypotheses in etiologic research. Spatial and space–time analysis should provide elements that foster the understanding and characterization of the processes that generate the spatial distribution of the phenomenon, based on the initial spatial distribution of various agents and risk factors. Furthermore, it should allow the phenomenon to be modeled by processes constrained by these characteristics.

This approach goes beyond the traditional "geographical" framework, as the matter at hand is less a characterization of the territory depending on its many components than the use of localization in order to highlight and test explicit hypotheses, using space as a common denominator. What space is and how it is used should be clarified: in the epidemiological analysis and in the search for explanatory factors, localization is often considered only by means of a classical statistical approach, which first involves the aggregation – in time and space – of (often mobile) elementary events in (immobile) geographical units that are subsequently analyzed.

The notions of spatial scale for description and analysis are essential, since they allow the synthesis of geographical units of knowledge based on individuals or elementary events. This knowledge is often available only in the form of probabilities, which are statistically estimated from observed situations.

As noted throughout this book, it is nevertheless difficult to conduct spatial analysis in epidemiology, particularly in etiologic research, as the spatial distribution of a phenomenon results from a set of entangled events and processes. Their "disentanglement" is the objective. Fortunately, it often happens that one of the processes carries the main "responsibility" for the spatial distribution of the phenomenon, and the study of this spatial distribution – after having eliminated the influence of already-known risk factors – is sufficient to formulate hypotheses on this process and its causal relationships with the studied phenomenon.

Several general remarks should be made:

– The number of individual events (changes in individual states such as infection cases) is part of the process analysis, but not of the actual spatial analysis.

EXAMPLE.– Spatial analysis allows the study of the spatial distribution of lightning impacts, but not of their overall number, which in spatial analysis is considered a constant parameter, equal to that of the observed situation.

– If an explanatory variable (a risk factor) has a spatial structure in terms of spatial relationships, but not of absolute localization, this spatial structure is found in the overall phenomenon. A spatial structure can be mobile: not every spatial structure necessarily emerges from an immobile environment.

EXAMPLE.– Lightning impacts in a storm are always close to one another, and have similar spatial characteristics, but storms are assumed to occur anywhere in the studied territory (Figure 2.1).

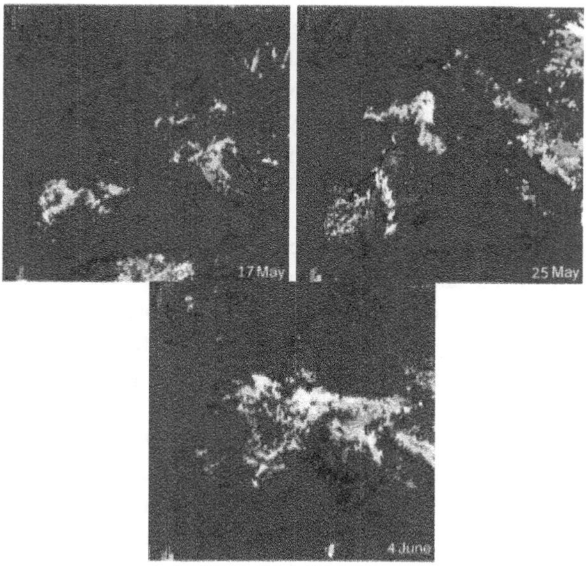

Figure 2.1. *Lightning impacts on 17 May, 25 May and 4 June 2018 in Western Europe. For a color version of this figure, see www.iste.co.uk/souris/epidemiology.zip*

– If individuals are spatially interconnected (for example, by proximity or encounter), the spatial distribution of the phenomenon will be influenced by these relationships, even in the absence of causal relationships.

– Some events or processes have a random spatial distribution (that is, due to many factors and/or chance); however, they have a crucial influence on the unfolding and spatial distribution of the overall phenomenon. It is often the case of low or very low probability events, whose position cannot be distinguished from a random situation (or of high spatial uncertainty phenomena, such as a rare meteorological event or the encounter between a vector and a pathogen). Due to these rare events, the overall system has high temporal or spatial variability (only necessary conditions can be stated). A separate study of the processes related to these rare events (their emergence, for example) is often required to separate them from the processes whose structure is less variable in space or time (for example, contagion and diffusion processes). However, an attempt will be made to quantify the probability of these rare events in the spatial distribution (by considering, if possible, longer periods of time for improved accuracies), or to evaluate the conditions for their occurrence, in order to apply the relevant precautionary principles.

EXAMPLE.– Statistical analysis shows that Ebola fever cases in Central Africa mainly emerge in hunter populations in forested areas, while no other environmental risk factor has been detected. It has therefore been concluded that its emergence is due to contact with forest animals infected by the virus (for example, monkeys), which were themselves infected when coming into contact with fruits contaminated by bats, the main potential reservoir of the virus. Diffusion involves other mechanisms (essentially the contact with an infected person, particularly during care or burial practices), which are less difficult to analyze.

– If, through a statistical study, a cause-and-effect relationship is observed between a risk factor and the studied phenomenon, the location of the events must *a priori* be considered a confounding factor: it is the presence of the risk factor in that place that is linked to the phenomenon, and not the place itself intrinsically. This is important since the presence of a risk factor in a place may be due to chance or temporary events: risk factors commonly have high spatial or temporal variability, which they also transfer to the health phenomenon. Identical repetition of two epidemics is very rare. Spatial and temporal variability of the risk factors detected through traditional statistical analysis must be studied before the characterization of places (and drawing risk maps, for example). *Direct characterization of places based on prevalence or observed risk before the detection of risk factors involved in the phenomenon and the study of their spatial and temporal variability should be avoided.* "Immobile" risk factors (place-related factors that allow their characterization in a geographical approach) should be distinguished from "mobile" risk factors, whose spatial variability must be studied and analyzed before place characterization. Risk maps whose construction relies solely on the localization of observed cases should be considered with caution!

– Studying the **emergence** of an infectious disease amounts to eliminating all the changes of states that depend on diffusion, and particularly processes linked to contagion and proximity between agents. If the study focuses only on emerging cases, which have *a priori* no direct spatial interrelationships, then statistical independence of individuals with respect to the sought-after relationship is increased. This is in principle a requirement in all the theoretical calculations of probabilities based on the binomial distribution and its normal or Poisson approximations. This restriction operation is similar to adjusting the data to a spatial character.

EXAMPLE.– If there is a high concentration of cases of an infectious disease solely due to proximity-based diffusion processes, this concentration may involve a high correlation between the number of cases and for example the type of habitat, while the phenomenon is not associated with the characteristics of the place where the concentration has been observed. Any cause and effect conclusion should be avoided. The cases are interrelated due to proximity, and are not independent.

– A random spatial distribution corresponds either to a phenomenon that has no link with localization (which is rare), or to a phenomenon whose multiple components are intertwined and lead to a spatial distribution that can no longer be distinguished form a random distribution. Fortunately, there is often a dominant factor in the phenomenon, and one part of the spatial distribution of this risk factor can be found in the spatial distribution of the phenomenon.

– In order to model a phenomenon after having studied the spatial distribution, a fully deterministic spatial process can be chosen, which will then be presented in the form of equations or rules whose variables are the assumed risk factors, with no random component (for example, a radial diffusion depending on the distance to a center). In this case, the random variability of the modeled phenomenon stems solely from the initial conditions (if they have a random component) and from the random variability of the (space–time) localization of risk factors.

Initial spatial analysis sometimes allows the formulation of etiologic hypotheses, but offers no information on whether the spatial characteristics of the phenomenon result from the spatial characteristics of one of its risk factors or of the intrinsic phenomenon characteristics (for example, case concentration due to contagion). Therefore, the analysis process involves the "unraveling" of the processes involved in the studied phenomenon. Detection of risk factors is a first step, allowing, in particular, the analysis followed by the elimination of intrinsic (non-spatial) risks and so-called "mobile" environmental risks before carrying out the spatial analysis for "immobile risks". Thus, proven risk factors can be "eliminated", and spatial analysis can be focused on residue and underlying phenomena. Once detected, risk factors can be analyzed as such, independently of each other (statistical distribution, spatial distribution and variability, spatial dependence) or all together (in a model) if a risk mapping is to be realized.

2.1.3. *Spatial and statistical dependence*

Localization is a source of complexity in a statistical study, as some hypotheses required by the statistical approach (for example, independence of the individuals in a sample, or absence of the correlation between values for unbiased estimation of variance) may be unverified – which is generally the case – when localization is willfully involved in the reasoning process and in the conclusions. Indeed, most of the events or factors are not independent in space, and an element that is implicitly used in the hypothesis should not be involved in the conclusion.

Spatial relationships exist in most natural or anthropic phenomena: objects are often interdependent according to their proximity, since the underlying phenomena depend on proximity; this is what geographers call "Tobler's law" or the "first law of geography" [TOB 70].

From a statistical point of view, we must be cautious and avoid concluding a dependence if a spatial relationship, regardless of its nature, exists and has been implicitly used in the reasoning or the choice of individuals retained for studying and quantifying the phenomenon.

For example, if a disease study is conducted on a group of villages, the spatial distribution of the villages must be considered and eliminated to avoid its involvement in the reasoning on the localization of cases.

Similarly, during the search for an explanatory statistical model based on risk factors, a component should be introduced to allow for considering the spatial dependence that is not explained by the cause and effect relationships with the risk factors (autoregressive models).

When a sample is drawn for a random survey, the elements must always be chosen independently, to avoid bias in the result. When the choice of elements relies on a spatial criterion, this assertion is no longer verified when the phenomenon presents a spatial dependence: variance within a disk increases with the disk radius. Any spatial conclusion must consider the spatial distribution of the sample if the latter has not been randomly chosen in space, or if the sample frame itself is not random in space. Furthermore, this choice often allows an improvement in the survey accuracy, through *a priori* use of the assumed spatial dependence.

Box 2.2. *Spatial relationships in health phenomena*

The study of spatial dependence will be revisited several times throughout this book, most notably in Chapters 5 and 6.

2.1.4. *Causal relationships, explanatory factors, confounding factors*

Localization is most often involved in a relative and non-absolute manner: from a purely epidemiological perspective, it can be considered a confounding factor. Indeed, absolute localization (longitude, latitude) will not be considered as such among the explanatory characteristics of a health phenomenon (except for some rare cosmic, astrological or meteorological phenomena). One or more other factors present in the same place or in proximity will be sought after as an explanatory cause. Highlighting the natural or anthropic constructions that allow the non-random occurrence of a phenomenon to be explained in a place is also one of the objectives of geography.

> In epidemiology, a confounding factor is defined as a factor that has a relationship with the studied phenomenon, but that hides, mitigates or reinforces the relationship with an explanatory risk factor of the phenomenon [BOU 93].

Box 2.3. *Confounding factor*

Several examples can be provided as follows:

– The study of the occurrence of lung cancer on a sample of the population shows a relationship between disease and gender: the number of male patients is larger than the number of female patients. However, when the sample is adjusted for smoking, the relationship with gender is no longer evident, since there are more male smokers than female smokers. In fact, the risk factor is smoking, and not the fact of being male: gender is a confounding factor.

– An area is characterized by high seismic hazard. Obviously, localization itself does not explain this hazard, but the contact between tectonic plates in that place does. On the other hand, highlighting places that are at risk advances the search for real causal factors. Furthermore, if the sole objective is spatial evidence of risk, and not finding an explanation that would reduce the risk cause, then localization should no longer be considered as a confounding factor. This is the common approach to seismic risks: since it is impossible to act on the drift of tectonic plates, and this drift is relatively slow, the place itself can be considered a causal factor.

– Suppose that rain facilitates the occurrence of disease cases. Space connects the two phenomena (disease and climate). If rain occurrence is independent of place (1), localization is a confounding factor: a specific location cannot be considered an explanatory risk factor, and drawing risk maps that rely exclusively on this factor should be avoided. If rain occurrence depends on place (2), cause–effect relationships can be reversed, and rain becomes a confounding factor (even though the phenomenon is caused by rain). It remains to understand why rain depends on place, and this may eliminate the place as a cause of the phenomenon, unless this meteorological phenomenon effectively depends on the absolute position.

(1) rain $\Rightarrow$ place $\Rightarrow$ case

(2) ? $\Rightarrow$ place $\Rightarrow$ rain $\Rightarrow$ case

2.1.5. *Uncertainty in event localization*

Geographical localization of agents (individuals, vectors, reservoirs, etc.) is not in itself an easy-to-measure parameter. Most agents are mobile, and their positions can be determined in advance only with very poor accuracy, if at all. The position of an agent may often be evaluated only using model-based probabilities (this is, for example, the case of wild animals – hosts, vectors or reservoirs). *A posteriori*, the position of patients (or non-patients), when available, is practically never provided with the accuracy of an individual spatio-temporal route. For individuals, it is most often predicted or provided only approximately or synthetically, with respect to a place of residence, work or hospitalization. It is very difficult to accurately evaluate an individual's real exposure to a hazard or risk factor. Once more, a model resulting from the statistical study of behaviors can be used as a tool to evaluate probabilities. Finally, when a health event occurs, it often happens that the place of residence or work does not coincide with the place of exposure or infection. This is especially true for pathologies that emerge long after an exposure or a contact (known as long-latency disease). Environmental factors, such as altitude or soil usage, are generally measured on immobile objects, even though their value changes in time (for example, rain, temperature, wind).

EXAMPLE.– It is common practice to analyze Dengue cases using patients' residential address, while the virus-carrying mosquitoes (mainly *Aedes aegypti*) bite essentially during the day, when most people are at work, at school, at the shops, etc.

2.1.6. *Health data are often aggregated into geographical units*

Health phenomena take place at the level of agents and relationships between agents (pathogens, reservoirs, vectors, hosts, patients, etc.), but it is often difficult, if not impossible, to obtain exhaustive information at this level of detail, for reasons of availability as much as for reasons of confidentiality and ethics [MAS 03]. An option may be to disregard the actual localization of agents, proceeding instead to their aggregation into immobile geographical objects with well-defined localization, such as areas resulting from administrative or geographical divisions. Most often, the studies on observed situations aggregate information into geographical units.

Spaces replace individuals as objects of study, and the whole analysis focuses on data aggregated in these units (counts, means, rates) rather than on system agents. It may be difficult to use counts or means if there is significant internal variation in these aggregation units, temporal relationships are neglected or aggregation leads to loss of information that is essential for understanding the phenomenon (for example, low probability events, or spatial or temporal proximity relationships between agents).

Aggregated data should be assessed in terms of quality and representativeness, by studying the distribution of values (particularly variance) in aggregation units. When the direct geometrical relationship between individuals and exposure to the risk factor is lost, any conclusion featuring this direct relationship, and in particular any causal relationship, is forbidden, being known as "ecological fallacy" (which involves the translation at the individual level of a correlation established at the group level). Finally, the use of aggregate data may prove delicate when several different processes are involved in the studied phenomenon (for example, emergence and diffusion: the use of cases issued from direct contagion by diffusion to evaluate the influence of the environment on the emergence may lead to false conclusions).

An approach that is *a priori* simpler than the spatial analysis of events is to evaluate the probability of a health event, not for individuals, but for spatial units (a place, a road section, an area, etc.), by extending the notion of a health event to a space. Risk factors are then regionalized random variables (their distribution of probability depends on place) corresponding to means, particularly for individual risk factors. This approach is known as spatial epidemiology.

Box 2.4. *Spatial epidemiology*

In all the cases, it is important to clearly specify the objects concerned by the analysis: the individuals or the objects of an aggregation. Hypotheses and probabilities must be expressed in terms of these objects (for example, the probability of being sick is used for individuals – which can be a function of the parameters of the individual – but for an aggregate object, several choices are possible: probability of observing at least one patient, or a proportion of patients, or a number of patients as a function of the population, density, average age, etc.).

Ecological fallacy is to assign to an individual a result that has been obtained for the group to which the individual belongs [GRA 94]. In fact, there are few cases where it can be said that what is true for a group is true for any individual of this group: this would require zero variance inside the group and the absence of any confounding factor at the individual level. We cannot reason in terms of probability, either, since group statistics do not generally correspond to the evaluation of probability concerning individuals.

A classic example is a study on the suicide rate conducted in the 1960s in Northern Ireland. District aggregate data show that the higher the proportion of Catholics, the higher the suicide rate. The conclusion that Catholics commit suicide more than the others is false: an individual study shows that the suicide rate among Protestants is higher. The reverse relationship observed at the district level can probably be explained by the pressure to which Protestants are subjected when they are in a large minority [VIG 97].

Box 2.5. *Ecological fallacy*

A contrario, aggregation in geographical units offers many advantages: if demographic data are available, it allows the calculation of rates, proportions and ratios in order to evaluate probabilities and risks, and it allows the use of other data characterizing the aggregation units. It also allows the passage from mobile objects (individuals) to immobile objects (places).

One of the objectives of the analysis of aggregated data is to detect units with "abnormal" values. This analysis of a set of units involves statistical techniques that will be detailed in Chapter 6. When statistical analysis does not lead to concluding the existence of an abnormal situation, further techniques should be used in order to prove that the observed situation is not due to chance, particularly by studying the exposure to a supposed risk factor. Localization is then used only to calculate the exposure. If individual-level analysis is possible, the accuracy in the localization of individuals should be sufficient to allow the calculation of exposure. If the analysis is conducted on geographical units and aggregated data, then the characterization of the exposure concerns the geographical units and not the individuals, and any causal connection between exposure and the health phenomenon at the individual level should be avoided. However, in all cases, other risk factors that may be involved in the phenomenon should be considered.

When individual data on the agents are available, aggregation can be *chosen*; it is not *inevitable*. This eliminates the constraints imposed by the shape or size of predefined units, which are not always adapted to the problem studied. As will be seen in the remainder of this book, many analysis methods use spatial aggregations in disks around a point in order to characterize this point: detection of clusters, local density or intensity assessment, rate assessment, smoothing and interpolation, detection of environmental correlations, etc.

2.2. Spatial analysis terminology and formalism

Before proceeding any further, and in order to formalize the above remarks, a review of the various types of objects, data and situations that are involved in spatial analysis, and that influence the choice of a specific method of analysis, is required. The language and concepts used for geographic information systems and for statistical and spatial analysis will be reviewed here.

2.2.1. *Objects, attributes, events*

– An **object** is an element described by attributes (or variables). A set of objects of the same type (whose description is based on the same attributes) is called a **collection** (known as a *relationship* or *table* in the “database” language, and a *layer* in the GIS language). As already seen in the introduction, a study often involves several collections of objects (for example, humans, patients, vectors, care centers, roads, weather stations, streams, areas describing the use of soil, departments, etc.)

– The term **agent** is often used for mobile objects, and the word **individual** for human or animal agents. The term (geographical) **unit** is often used for immobile objects, such as villages or administrative areas.

– The objects in a collection are described and characterized by **attributes** (also known as **variables**). These attributes can be **quantitative** (numerical values, measurements or quantities – counts, integer values – or ratios – frequencies, percentages, densities, etc.) or **qualitative** (nominal values that are finite in number, such as names of places or colors). Qualitative values can be **Boolean** (true/false). A Boolean value also allows the definition of two subsets (corresponding to the two values *true* and *false*), and allows us to have a set-based approach in spatial analysis (set membership, intersection, etc.).

Examples of numeric attributes: a temperature measurement, an altitude measurement, number of vehicles per hour, age, size, weight, number of inhabitants, number of cases, density, incidence, etc.

Examples of qualitative attributes: an identifier, a type of soil, a profession, a medical status, the severity class of an infection, an exposure, a type of bird and any qualitative classification of a quantitative variable (a class of age, a class of altitude, etc.).

Examples of Boolean attributes: true/false, patient/non-patient, present/absent, etc.

Box 2.6. *Various types of attributes: examples*

– An **event** is defined as that which changes the value of an attribute of an object. Most often, a date and/or a position are associated with this change: the event is then considered **localized** (in space, time or both). Events are thus objects which are also described by attributes and grouped into collections.

– It often happens that lists (collections) include only those individuals that feature an event that characterizes them (for example, a registry of patients, with a date of hospital admission or a date of infection). The individual and the event are then confounded. If the event is localized (in time and/or space), this location is considered to be the localization of the individual. (In fact, the analysis concerns events, and not individuals. The attributes of the individual are added to the attributes of the event.)

2.2.2. *Localization and spatial domain*

Throughout the remainder of this book, an object is said to be **localized** in space if it has a geographical location in a well-defined system of coordinates (geodesic system, geographical projection, etc.). Location is a two- or three-dimensional attribute in a Euclidian space, which distinguishes it from "classical" one-dimensional attributes.

When objects (individuals or events) are localized, the geographical set to which the studied objects may belong is called the **domain of study** or the **domain of definition**, which is denoted by **D** (**D** $\subseteq \mathbb{R}^2$ **or** $\mathbb{R}^3$). This domain can be a continuous set (two- or three-dimensional space domain), a continuous subset, said of dimension 1 (for example, a straight line, a curve or a network), or a discrete and finite subset (for example, a set of points or areas, considered as homogeneous units). The domain D can formally be defined by a probability of occurrence p given at each point of the space ($D = \{P \in \mathbb{R}^2 \text{ such that } p(P) > 0\}$). When time is also available, the objects are said to be localized in space and time; the domain D is then included in $\mathbb{R}^2 \times \mathbb{R}^+$ **or** $\mathbb{R}^3 \times \mathbb{R}^+$.

In a study of lightning impacts, the domain of study is considered continuous, as the events can occur anywhere on Earth. The domain is also continuous in a study of earthquakes; thanks to geological studies, a probability of occurrence can be assigned to each point of space and considered in the analyses. For a study of road accidents, the domain is the road network, which is also considered as a continuous one-dimensional space: an accident can occur anywhere in the road network, but only in this network. If a survey on a river water quality is conducted, the domain is only the river, considered either as a surface or a curved line, depending on the required accuracy. In a study on the localization of cases of a disease, located by belonging to a village, the domain of study is the set of villages, considered as points: the set of possible places is finite, not continuous (in mathematical terms, it is **discrete**).

Box 2.7. *Continuous domain or discrete domain*

Spatial analyses that can be conducted depend on these different situations, as listed below:

– When the domain D is continuous, it is possible to study the absolute position of a set of events in this space as such, independently of an arbitrary descriptive attribute (for example, the position of lightning impacts, the position of earthquakes, car accidents, a tree species in a forest). The whole set of events should be known, and not only the events belonging to an *a priori* defined subset (such as the measuring stations), in which case the domain D is reduced to this discrete subset.

– When the domain D is continuous and the study focuses on the spatial distribution of the value of an attribute rather than the localization of a set of objects, this value is supposed to be present everywhere and can be measured at any point (for example, temperature, altitude, type of soil). The set of studied objects constitutes the set of all points of the surface (in fact, a geographic accuracy is always implicitly defined, a spatial resolution that involves a finite number of points, as on a grid). If the attribute is measured only on a well-defined subset (weather stations, topographic surveys, etc.), then interpolation methods should be used to estimate it at any point of the space, or the study should be limited to a finite and discrete subset of measuring stations.

– When the domain D is discrete, the analysis of the position of the studied set (subset or value) is more difficult to apprehend, since it must take into account the absolute position of D. In general, this absolute position is not part of the study. Only the relative position of the values of an attribute is studied, by considering the absolute position of the set (for example, if D represents a set of villages, the goal is to study the position of the infected villages with respect to the position of the set of villages; if we study the clustering of infected villages, the calculation will consider the clustering of villages, which is not part of the study).

– When the domain D is discrete, a qualitative attribute allows the definition of a subset in the studied set. Then, the position of the subset defined by one of the modalities is studied, still taking into account the characteristics of the spatial distribution of the studied set.

The term **point cloud** is applicable when the localization of an object can be spatially reduced to a point. The objects are then spatially represented by points. The set of objects constitute a "point cloud".

For example, in a scale analysis of a region or a country, a set of villages is often represented by a point cloud. The presence of a disease in some villages allows the definition of a subset of this point cloud dataset.

The presence of a tree species in a forest can also be represented by a point cloud, each point corresponding to a tree of the species considered, in the middle of a space where the other trees are not known, but they are assumed to fill all the space or to be randomly, spatially distributed in this space.

Box 2.8. *Point cloud*

– As already mentioned (and as will be revisited in Chapters 3 and 6), the individual epidemiological data (the events) are often aggregated in spatial units. These units often represent a surface (for example, administrative units or unit cells). Therefore, the nature of attributes changes, and they no longer describe individuals, but populations (the set of individuals of the unit). The value assigned to a spatial unit can be a mean, a count or a frequency calculated from the individuals belonging to the spatial unit. In terms of spatial analysis, this leads to a domain D formed of a finite number of objects (the spatial units), whose localization is often schematized by a point (called a centroid). Aggregation leads to significant loss of spatial information, notably the accurate localization of individuals and the spatial relationships between individuals. In contrast, aggregation has the advantage of transforming mobile objects into immobile objects that are much easier to graphically represent and analyze. A further advantage – or drawback – of aggregation is that it allows the calculation of means, counts, frequencies and densities, thus reducing statistical variability.

2.2.3. *The formalism of descriptive analysis*

Let us consider a population of individuals belonging to a domain D. The phenomenon under study corresponds to the value $\boldsymbol{y}$ of an attribute of these individuals or of an event concerning these individuals (for example, a morphological characteristic, an immune status, the emergence of a disease, etc.). When the objects or the events are localized, this set presents a spatial distribution, and the existence of a mathematical expression F can be assumed, depending only on space–time localization, which enables the mathematical description of the values $\boldsymbol{y}$ observed at instant t and provides information on the phenomenon:

$$\boldsymbol{P,t} \overset{\boldsymbol{F}}{\rightarrow} \boldsymbol{y} = \boldsymbol{F(P,t)}$$

As already noted, the phenomena observed are due to many processes, spatial and non-spatial. Therefore, they often exhibit high spatial variability, at all the geographical scales used for their observation. In practice, the variations of $\boldsymbol{F}$ as a function of localization P are highly irregular. An analytical expression linking position and value would be extremely complex, if not impossible, to establish. The variations of function $\boldsymbol{F}$ could even be purely random if localization has no relationship with the studied phenomenon; however, this rarely happens, as already noted: in contrast, it frequently happens that the function ***F*** presents structured variations or non-random global tendencies, which is precisely what we are trying to highlight. Even if this *a posteriori* descriptive analytical expression were determined as a function only of localization, it would nevertheless be of little interest in the explanatory research of mechanisms, since it only describes the spatial distribution of the observed phenomenon, and involves neither the characteristics of objects or processes, nor the initial conditions.

In epidemiology, the main goal of descriptive spatial analysis is to determine the characteristics of this function ***F*** and to eliminate the spatial noise. In other words, descriptive spatial analysis aims to identify the characteristics and spatial tendencies of the processes: overall characteristics of the observed spatial distribution (random, dispersed, aggregated), extent, magnitude of local variations, study of spatial continuity and autocorrelation, detection of distinctive places, identification of spatial tendency or global pattern. The function ***F*** is only known for a finite number of individuals, but because of the two dimensions, the mathematical tools that allow the characterization of geometric parameters are much more numerous than in one dimension.

One of the main subjects of spatial analysis is the variance of ***F*** for the points of the domain D situated within a disk O of radius R around a point P, and the study of variations of this variance when R increases or when the point P varies in D. The variance can be that of a numeric value (process valued by a numeric variable), a frequency (process valued by a Boolean variable) or the number of events in O (non-valued punctual process).

In order to manage and exclude the random spatial variability (noise) due in particular to non-spatial factors and processes, a probabilistic approach is certainly required for this function ***F*** and its spatial variations. The value observed for each localization P is therefore seen as the single realization of a random variable ***Y*** (a set of possible values subjected to a distribution law of probabilities). The notion of a random variable is then extended to the function ***F***, which is considered as a field of random variables on domain D, called a regionalized variable, assuming that the phenomenon is identical throughout D, meaning that all the random variables ***Y*** follow the same laws. The study of spatial characteristics, variations and tendencies of ***F*** is conducted within this probabilistic context. Studying the spatial dependency of ***F*** amounts to studying the characteristics and spatial dependency of the random variables that compose it. When the domain D is continuous, it is the object of geostatistics.

Geostatistics is a branch of statistics that integrates the localization of random variables and aims to study the processes that unfold continuously in space. Its main applications are in geology, climatology, telecommunications, etc. Geostatistics uses exploration tools (variogram, correlogram), estimation and interpolation (kriging) techniques and simulation techniques.

Box 2.9. *Geostatistics*

A **spatial point process** is a stochastic process in which the localization of events is itself considered as a random variable. A realization of a spatial process constitutes a set of localizations generated by the process in time. There is a distinction between **non-valued** processes (the event has no attached value, only its position is taken into account and the function ***F*** is Boolean) and **valued** processes (besides position, a one-dimensional numeric random variable is linked to the event).

The objective of the descriptive spatial analysis of a spatial point process is to identify the mechanism that generates its spatial distribution. The analysis of non-valued spatial processes relies only on the number of events N(A) for any set A in a region of study D. Spatial processes have been the object of numerous theoretical developments aimed at their modeling. There are many theoretical models, such as:

– **homogeneous Poisson** process: this corresponds to a completely random situation, which allows the modeling of a spatial point process in which the occurrence probability of an event is the same in any point, and in which events are independent;

– **non-homogeneous Poisson** process: mean density of events may vary in space while preserving the independence between events and the randomness of the number of events;

– **Neyman–Scott** process: each first-generation event generates a point process in its vicinity following the same law as the initial process and independently of other events;

– **Matern** process: each first-generation event generates a random number of events that are uniformly distributed in a disk of radius R;

– **Thomas** process: each first-generation event generates a random number of events whose distribution is Gaussian and isotropic around the event;

– **Cox** process: non-homogeneous Poisson process where local density is considered a variable of time;

– **Markov, Gibbs** process: modeling of processes in which events are not independent, such as contagion/inhibition processes – the occurrence of an event at a point increases (for contagion) or decreases (for inhibition) the probability of occurrence of other events in the proximity of this point.

Most of these models relate to applications for which the domain D is continuous, and in which the events can occur or be measured at any point, which is seldom the case in the health field. Therefore, they have to be adapted to a discrete space, with probabilities of occurrence of the events that depend on one or several variables (population, susceptibility, risk factors, etc.).

Box 2.10. *Modeling of non-valued spatial processes*

In epidemiology, at the finest spatial scale, the domain D is discrete or one-dimensional: it is the spatial support of the phenomenon, as a set of individuals (located at instant t), a set of residences or farms, a set of villages, a road network, etc. The processes to be analyzed concern the distribution – notably spatial – of a value (quantitative or qualitative) in the spatial support and not the localization of the spatial support or of the events themselves. The modeling of these processes cannot make use of the above-mentioned theoretical models, which are valid only in the continuous case. In general, the analysis methods absolutely must consider the characteristics of the spatial support.

The actual distribution of the spatial support or of the events is most often ignored and localization greatly simplified: this is particularly the case for data aggregated per geographical unit, which do not consider the spatial support or the actual position of events, but only their belonging (geometric or relational) to a spatial unit. The actual distribution of the spatial support is ignored, and for modeling purposes, the units are considered to be continuous spaces. The spatial distribution of the support is replaced by an intensity or a probability in each unit. The spatial phenomenon can then be modeled as if it were continuous within each spatial unit, and continuous models can be used for spatial processes. For example, for geometrically aggregated data, the random situation can be modeled by a spatial non-homogeneous Poisson process, assigning to each geographical unit its own intensity (generally corresponding to the evaluation of a risk factor in the unit, for example a density of population at risk).

2.2.4. *The formalism of the explanatory analysis*

Epidemiology goes beyond a descriptive approach, attempting to also be explanatory. Given the systemic approach already described, one of the main objectives of scientific health research (and of epidemiology in particular) is to determine the variables of this system and construct a mathematical expression that gives the probability of the health event for each individual (a change of state for a health parameter of the individual), wherever it is, depending on these variables (individual risk factors, environmental risk factors, mobile or immobile, vulnerability, interactions within individuals, etc.):

QUESTION.– Find a mathematical expression for the law of probability of a random variable or of an event using individual, environmental or contextual variables.

Most risk factors must themselves be considered as regionalized random variables (an unknown value whose distribution of probabilities, which may depend on place, must be determined). Quite often, only the mean is taken into account, particularly for individual risk factors. All of the explanatory analyses of health geography attempt to explain the influence of these factors and contexts, and to understand their mechanisms.

Formally, given an individual P and a studied attribute y (for example, a disease risk for the individual), the objective is to estimate y as a function of predictive variables $x_1, x_2, \dots, x_n$ (risk factors). Ideally, the goal is to find a model M, allowing the estimation of y:

$$x_1, x_2, \dots, x_n, P, t \overset{M}{\rightarrow} y = M(x_1, x_2, \dots, x_n, P, t, \varepsilon)$$

The process M is considered as identical throughout the domain D, and the observed spatial variations are due to variations of risk factors, interactions between individuals or random variations. The expression of the model M is *a priori* unknown, and thus variables x_i should be taken into account in the model. Localization is involved when variables x_i are localized (for example, environmental variables), when the absolute localization P is itself a risk factor (which is rare, as already mentioned) or when the phenomenon depends on the interactions between individuals (for example, contagion/inhibition processes or conjunction processes). Interactions can themselves depend on localization, for example the places where the probability of contact is higher (for example, markets, schools, public transportation, stadiums, etc.).

The model M is often too complex to be determined: it is preferable to consider the studied phenomenon as the sum of various independent processes $M1, M2$ ... Mn, which are *a priori* simpler than the overall model (for example, by separating the emergence processes and the diffusion processes, vector-related processes and individual-related processes, etc.). A preliminary analysis of the processes involved in the overall phenomenon is therefore important. It allows the formalization of the description in terms of a complex system as has been explained earlier in this book.

Therefore, the approach differs from the descriptive approach. The expression of these independent processes Mi requires their reasoned construction (choice of risk factors, choice of parameters), sometimes independently of the values observed during the complete process. Each process Mi can be submitted to an individual statistical and/or descriptive spatial analysis based on the observed data, in order to find an analytical expression that models it.

Based on the observation of an actual situation, residue analysis is conducted after successive elimination of Mi, in order to reach a residue that is, in principle, purely random. When variable y is localized, the residue, itself localized, should correspond to the spatial noise (due to spatially random variables and purely random spatial variations), and should have no spatial structure. If this is not the case, this means that models Mi are not adequate: either their expression is not satisfactory, or a process has been omitted in the overall model. This type of analysis involves many techniques, particularly techniques for statistical modeling of spatial data. It also involves process simulation in order to evaluate the distributions of probabilities of the random variables studied.

The generalized linear models (for example, the logistic model) are among the most used for statistical modeling in health, as they often correspond to what is already known on the pattern of the process to be modeled (and whose coefficients are to be determined).

When individuals and risk factors are localized, the classical statistical models only consider the relationships between the variable to be explained and the explanatory variables, while omitting spatial relationships between individuals. It often happens that the residue is not randomly distributed in space and features spatial autocorrelation. **Autoregressive spatial models** have therefore been developed to integrate this autocorrelation in the model without necessarily trying to explain it. Their expression has the following form:

$$X, P, t \overset{M}{\rightarrow} Y = M(WY, X, P, t, \varepsilon, W\varepsilon)$$

X represents the vector $x_1, x_2, \dots, x_k$ at point P (the values of k risk factors). Y represents the vector $y_1, y_2, \dots, y_n$ of values observed for N objects. W is $N \times N$ **matrix of spatial weights** between objects. It is a square symmetric positive matrix, built from topological or metric relationships between the objects: adjacency of order 1 to n, functions of the distance between objects (for example, $(d0 - d)/d0$ or $(\frac{d0-d}{d0})^2$, etc.) and variance minimization. This matrix of spatial weights is present in many calculations in geostatistics and spatial modeling (see Chapters 5 and 6).

Considering that the model itself can vary in space, at least in terms of its parameters, it is possible to attempt to adjust these parameters (the coefficients of the model) depending on the localization, taking into account the local characteristics of spatial dependency in the resolution of the model. It is thus possible to build models Mi (particularly contagion models) using the study of local spatial characteristics of the overall residue observed before taking into account the spatial relationships between individuals (continuity, autocorrelation, pattern, etc.). These **models of regression with a spatial weight** (GWR, see Chapter 6) have the following expression:

$$X, P, t \overset{M}{\rightarrow} Y = M(X, WX, P, t, \varepsilon, W\varepsilon)$$

Another approach is that of **multi-agent modeling**, which avoids searching for an analytical expression calibrated solely on observed data and thus extends the scope of possible situations. Moreover, the definition of agents, their behaviors and interactions allows the modeler to be much closer to the initial systemic model. Finally, it can be used to directly elaborate spatial–temporal models. However, the determination and calibration of behaviors remain difficult tasks, and so does the validation of the model, which often faces a single realization of the phenomenon.

2.3. General approach of spatial analysis in epidemiology

Similar to classical statistics, spatial analysis can be descriptive or inferential when it attempts to show cause-and-effect relationships.

Descriptive spatial analysis seeks to characterize F, the spatial distribution of the phenomenon or of one of the Mi processes. Explanatory spatial analysis allows the elaboration of hypotheses on the phenomenon and its explanatory factors. The objects analyzed are either individuals with their characteristics or objects with counts (corresponding to the aggregation of individuals) or characteristics (mean values, environmental descriptors, etc.), or the residue of a model (the difference between the observed values and the model). The spatial analysis methods to be used depend on the aggregation level of the available data.

At each stage of the analysis, mapping allows a visual interpretation of the results. These visual interpretations do not stand as proofs, as a statistical test would; however, they allow the formulation of hypotheses (to be refuted or validated by statistical tests), or can be used as support for a geographic argument.

2.3.1. *The approach of descriptive analysis*

Descriptive analysis of spatial distribution is part of the more general approach of descriptive analysis of epidemiologic data. It involves the following:

– Study the overall characteristics of the set of individuals (means, frequencies, etc.), independently of the localization or aggregation level of data, if there is any, in order to estimate as accurately as possible probabilities on the set of individuals. This is the classical approach of descriptive epidemiology.

> Whatever the type of data, the first analyses concern the estimation of epidemiologic indices based on counts observed on the entire studied population (prevalence, incidence, risks). These statistic calculations concern overall counts and can be made based on individual or aggregate data, since only counts are used.

Box 2.11. *Using spatial analysis in addition to classical statistical analysis*

– Conduct a first visual analysis: mapping and graphic representation of the observed situation, trend surface (see Chapter 4, section 4.4.2). This is a first geographical approach, which allows an overall vision of the phenomenon and its spatial tendencies. It depends on the type of data, which is explained in Chapter 4. It is worth noting that, as it will be seen in what follows, visual analysis mixes several aspects (significance of values, spatial clustering, spatial tendencies) and has no *proof* value.

– Conduct a classic temporal analysis (tendency, cycle, seasonality, residual variability) when the data include a date.

– Aggregate individuals according to a qualitative attribute, by aggregation into objects (spatial or non-spatial), and study the characteristics of these groups and the differences in values between these groups, according to a classical statistical approach. If the objects are individuals, these individuals can be aggregated into groups depending on the availability of an aggregation attribute. The study concerns each group (estimation of risks in each group) and comparison between groups. A particular goal will be to estimate the significance of differences between groups or with respect to an *a priori* distribution (in most cases, a random distribution). The groups can rely on a classic definition in epidemiology (case–control, exposed–unexposed, gender, age group, etc.) or on a spatial definition (belonging to a place, such as a unit cell or a geographical unit). This aggregation process[1] must be well controlled, particularly when there are many groups: unit variance increases when the number of groups increases. From a certain level of disaggregation onwards, the distribution of values in the groups could no longer be distinguished from a uniform random distribution. When groups are based on localization, the characteristics of their spatial distribution could also be studied (gradient, continuity, direction, pattern, etc.), as indicated below.

1 From a statistical perspective, the process of aggregation of individuals into groups is rather a process of disaggregation of the studied group: the set of individuals – that was a single group – is broken into distinct subgroups, which will be analyzed separately.

If the data represent counts and are already aggregated into groups, only higher hierarchical levels of aggregation can be used.

When accurate localization is available at the level of the studied objects (individuals or geographical units):

– Characterize the overall spatial distribution of the events by synthesis indices on the absolute position, on the spatial layout of the objects or their values (aggregation/dispersion, spatial dependency, variogram, measurement of spatial autocorrelation) and on the overall pattern of the phenomenon.

EXAMPLE.– A possible objective is to characterize the intensity of a process of diffusion based on contagion (mean, variance). These characteristics can be assumed to be independent of place if local characteristics have no influence on the diffusion process.

– Identify the characteristics of the overall phenomenon (overall tendency, pattern), and a theoretical spatial distribution or a process allowing the modeling of the observed spatial distribution.

– Find distinctive places (source centers and places, clusters, exclusions, hot spots, cold spots). Study the spatial relationships at the individual level.

– If time is available, conduct spatio-temporal analyses: find index cases, path reconstruction, cluster emergence – movement – disappearance, diffusion models, extinction models.

When data are available at the level of individuals and the latter are accurately localized (in time and space), all the statistical, spatial and spatio-temporal analyses are possible.

Box 2.12. *Localization of the individuals allows many analyses*

2.3.2. *The approach of explanatory analysis*

Explanatory analysis concerns the study of risk factors and processes of emergence, diffusion and extinction. The main objective of classical statistical analyses is to study the relationships between the attributes characterizing the individuals, in order to model the influence of each variable on the studied phenomenon. The final objective is to model an individual risk based on the attributes considered as risk factors. These attributes can stem from the initial data themselves (description of the individuals: age, gender, physical and biological characteristics, socio-economic characteristics, cultural characteristics, etc.), from the interaction between the individuals (spatial relationships of proximity, contacts, contagion, etc.) or from the interaction with the environment (belonging to a space, context, influence of a density, probability of the presence of a pathogen, probability of the presence of a vector, probability of contacts, etc.). Geographic information systems (GIS) are very useful for the calculation of these attributes if individuals and risk factors are localized. When the individuals are localized, it is also necessary to complete these analyses by taking into account the spatial relationships between individuals: study of the local spatial characteristics of the phenomenon (continuity, autocorrelation, pattern, etc.) and consideration of the neighbors in the risk modeling process.

Environmental variables, which often have a non-random distribution, are used as characteristics of individuals, which are obtained either by direct collection or by belonging or aggregation calculation (using, in particular, geographic information systems). Classical statistical methods are essentially used to analyze the influence of these attributes on the studied phenomenon, at the level of individuals or groups of individuals.

The explanatory analysis seeks to determine a model, the theoretical forms of which have been described in section 2.2.4. Spatial analysis must then be conducted on the residue of the model. If it features a non-random spatial distribution, the model should be completed by considering the spatial relationships between individuals, either by introducing a component of spatial autoregression (a regression of the observed values with observed values of other individuals) or by adopting a model with spatial variations.

2.3.3. *Spatial analysis methods*

Many methods allow the introduction of localization into data analysis. Beyond cartography, data analysis methods can be classified according to the purpose of the study. They can be classified into overall analyses (overall characteristics of the spatial distribution on the entire domain of definition), allowing the evaluation of the overall spatial tendency, and local analyses (local characteristics of the spatial distribution and spatial interactions), allowing the evaluation of local spatial variations. They can also be classified into analyses of absolute positions and analyses of relative positions.

The methods include:

– overall analyses, known as *first-order analyses*, which provide information on the characteristics of the overall spatial distribution (autocorrelation, heterogeneity, overdispersion, tendency, centrality, pattern);

– local analyses, known as *second-order analyses*, which provide information on the local variations of the spatial distribution (excessive risk, clusters, hot spots, cold spots);

– detection of places in space (points or surfaces) that have a specific characteristic with respect to the spatial distribution of objects (cluster, centrality, attraction, repulsion);

– estimation and interpolation, which allow the characterization of any point depending on the spatial distribution of objects or on their values;

– process analysis (emergence, diffusion, extinction) using spatio-temporal modeling.

Analysis methods can also be classified into two groups:

– direct methods, which directly use the localization of objects considered as points directly in the calculations, the set of objects forming a *cloud of points*;

– indirect methods, which use a classical statistical analysis after aggregation or belonging calculation in a geographic division or a regular grid of space (which can differ from the geographic division of data aggregation when it already exists). The geographic aggregation process is thus at the core of spatial analysis.

> Analysis methods depend on the type of spatial implantation: they rely either on the geographic coordinates of the objects, and use the localization of objects directly in the calculations (direct methods), or on the aggregation of objects into spatial units (indirect methods, statistical analyses after aggregation).
>
> Direct methods can also be used with aggregated data that have already been analyzed with indirect methods: it is a matter of scale description and analysis. Loss of information due to the process of data aggregation into geographical units should nevertheless be taken into account.

Box 2.13. *Direct and indirect methods*

2.3.4. *Spatial analysis and health geography*

The use of spatial analysis in health geography particularly addresses the problems of health inequalities, the analysis of the healthcare system, the problems of supply and demand matching, and traffic accidents and road safety.

Accessibility is an essential factor in the analysis of the use of the healthcare system. The objective of spatial analysis is to observe and analyze the socio-territorial health inequalities in a territory. Healthcare accessibility is not only a spatial concept: it is a multidimensional concept that covers spatial aspects (healthcare provision, presence of equipment, transportation, etc.) and non-spatial aspects (socio-economic environment, cultural environment, etc.).

Many indicators have been developed to measure the spatial accessibility of a group of users to a set of health services or professionals. Spatial analysis is involved in considering spatial characteristics and in the definition of these indicators: patient mobility, transportation resources, route valuation, patient lists, resource allocation, etc.

The spatial analysis methods used in this context are not specific to health. They also apply to subjects such as traffic analysis (frequency of passage or time staying in a place), analysis of attractiveness, analysis of transportation networks, analysis of movements. As always, the first stage of work, which is essential, involves the clear specification of the objects to be studied.

2.4. Required knowledge on epidemiology and statistics

Proper use of spatial analysis tools requires knowledge on the general principles in statistics. Statistics is also omnipresent in epidemiology. This section is a basic review of the notions that will be used in this book. It is not intended as a substitute for reading comprehensive works on statistical methods used in epidemiology (for example, [BOU 93, ANC 02]). Detailed presentation of some statistical methods will be given in what follows, particularly those that are rendered more complex when localization is considered.

2.4.1. *Epidemiology*

Knowledge of the main notions used in epidemiology is essential:

– notions of bias and confusion, as have been mentioned earlier in this book;

– the principle of epidemiologic studies: case–control studies, cohort studies;

– notions of risk and its assessment based on counts and rates calculated over data observed and aggregated into geographical units: incidence, prevalence, survival; relative risk, *odds ratio*, attributable risk;

– notions of direct and indirect standardization: standardized incidence ratios, standardized morbidity or mortality ratios (SIR, SMR). These will be revisited in Chapter 6;

– statistical modeling of risk: generalized linear regressions, logistic regression, Poisson regression, etc.

2.4.2. *Statistical analysis*

Epidemiology intensively uses a number of statistical methods of which the principles must be mastered.

> In epidemiology, the most frequently used classical statistical methods concern:
>
> – the calculation of statistical characteristics (moments, significance) and case-related indices (prevalences, incidences);
>
> – the study of differences between groups, particularly in studies of exposure to a risk factor (case–control, cohorts);
>
> – the study of differences between groups for the study of an exposure factor (exposed vs. unexposed);
>
> – modeling of disease risk as a function of risk factors;
>
> – methods for the surveys.

Box 2.14. *Statistical analysis*

2.4.2.1. *Principles*

Statistics relies on the theory and calculation of probabilities, the main objective of which is to assess the probabilities or the models based on observed data. In general, the purpose of statistical analysis is to study:

– characteristics of the distribution of an attribute of the individuals in a population: mean, variance, mode, frequencies, skewness, kurtosis (moments);

– relationship of an attribute with one or more other attributes of the individuals in a population (covariances, correlations);

– aggregation of individuals into several groups depending on their characteristics;

– comparison between groups of individuals depending on their values (comparison of means, variances, distributions);

– value or risk modeling depending on attributes.

Statistics is either descriptive or inferential (when its goal is to provide probabilities on causal relationships, A $\Rightarrow$ B. Assertion B generally corresponds to an attribute to be explained, z, and assertion A to explanatory attributes, $x1, x2, \ldots, xn$). Statistics can be "frequentist" (based on the frequency of random events and the law of large numbers) or "Bayesian" (seeks to assess uncertainty and relies on Bayes' theorem on conditional probabilities).

Numerous methods are used to study the relationships between the attributes of an individual. The simplest method is to search for the correlations between two attributes, namely joint variations of two attributes. Indices built on the mean of the product of two attributes are used, expressing the multiplicative effect that one can have on the other. Covariance matrices of a set of attributes considered as random variables are used in multivariate analysis. Diagonalization of these matrices makes it possible to determine the main components, representing the reference corresponding to the linear combinations of maximal variances of initial variables.

The general objective of statistical modeling is to find an expression of the type $z = f(x1, x2, \ldots, xn, \varepsilon)$, allowing the modeling of variable z using a number of variables xi and a random factor ε (statistical noise), and to assess the variations between this theoretical expression and the values actually observed.

> The simplest method seeks to determine a linear relationship between one variable and another ($z = ax + b$). In the case of a multivariate problem, the form of the expression is $z = a0 + a1x1 + a2x2 + \cdots + anxn$. Sometimes modeling does not concern an observed attribute, but an expression derived from this attribute (for example, logistic regression, which seeks to find a linear expression as a model of $\ln(\frac{p}{1-p})$ from an observed probability p). Coefficients $a0, a1, \ldots, an$ are determined from the observed data using classical data analysis methods (minimizing mean squared error, maximizing likelihood).

Box 2.15. *Linear relationships*

2.4.2.2. *Confidence intervals*

The individuals observed and analyzed must be considered as a sample of the population to which they belong. Moreover, the values measured are always affected by measurement errors or accuracy uncertainties. The observed values and the calculated indices must be considered as approximations of accurate actual values (assumed to exist), unknown and submitted to sampling fluctuations and measurement errors. The estimation of these fluctuations and errors allows the definition of a confidence interval, which contains the actual value with a given probability, set by the user.

The entire difficulty lies in the assessment of these sampling fluctuations and measurement errors. Sometimes this assessment can be purely mathematical, for example when the choice of individuals is fully controlled and the measurement is not subjected to any error. Most often, a hypothesis is formulated on this distribution, estimating, for example, that it is due to chance or that it follows a given law of probability. The calculation of probabilities can then be used to calculate the confidence interval.

EXAMPLE.– Let us take the coin toss game as an example. There is no information on whether the coin is biased. The goal is to determine the probability p of obtaining tails when flipping this coin, in order to compare it with the theoretical probability obtained with a fair coin (0.5). For this purpose, the coin is flipped n times, and tails show k times.

Supposing that tosses are independent, the theoretical probability of obtaining these n results is given by the following binomial law:

$$P(n,k) = \binom{n}{k} p^k (1-p)^{n-k}$$

The mathematical expectation of the binomial law is equal to np, and assuming that the observed value k is equal to the expectation, $p_{obs} = k/n$. However, the value of k is subjected to random fluctuations, which cannot be considered by this calculation. The law of large numbers makes it possible to estimate the distribution of the values of k and to give to the true value of p a confidence interval I around p_{obs}, with a risk α of making a mistake:

$I = \left[p_{obs} - z_{\frac{\alpha}{2}} \frac{\sqrt{p(1-p)}}{\sqrt{n}}, p_{obs} + z_{\frac{\alpha}{2}} \frac{\sqrt{p(1-p)}}{\sqrt{n}}\right]$, where $z_{\frac{\alpha}{2}}$ is the normal reduced-centered value.

This also makes it possible to verify whether 0.5, the theoretical value of p for a fair coin, is in this interval or not. It can be noted that the larger the value of n, the narrower the interval, and the closer p_{obs} will get to p.

2.4.2.3. *Principle of statistical tests*

Before any study of relationships between variables and any attempt at inferential modeling, many scientific questions can be expressed in the form of a hypothesis to be refuted or not. The role of statistics is thus to assess the probability of this hypothesis to be true by assessing the possibility that the observed situation is solely due to chance: it is the principle of statistical tests.

In order to answer a scientific question, a null hypothesis denoted H0 (for example, x = a) and an alternative hypothesis denoted HA or H1 (for example, x > a) are formulated. The hypotheses use an index (a numeric measure) that synthesizes the problem to be addressed. The test involves:

– studying the theoretical distribution of this index in the population, by using the theory and the calculation of probabilities or a simulation process. The theoretical distribution corresponds to situations that vary randomly, through draws from a supposed distribution of probabilities, generally equiprobability;

– comparing the alternative hypothesis to the theoretical distribution of the index, in order to estimate the probability of this hypothesis;

– rejecting or not the null hypothesis, depending on a risk of error;

Type I error (or α risk) is the probability of refuting the null hypothesis when it is true. Type II error (or β risk) is the probability of not refuting the null hypothesis when it is false. The power of a test is 1–β.

Box 2.16. *The principle of a statistical test*

EXAMPLE.– A pH measurement yields 6.7. The null hypothesis is H0: pH = 7, and the alternative hypothesis is HA: pH < 7. A risk of error of 0.05 is chosen. The study of the distribution of random measurements shows that the probability of obtaining 6.7 by chance while the value is 7 is below 0.001, far below 0.05. Therefore, the hypothesis that pH = 7 can be rejected, and it can be admitted that pH < 7.

The theory of probabilities is often used to calculate the random theoretical distribution of the values of an index. This evaluation is sometimes too complex from a mathematical perspective. A simulation can thus be used to evaluate this distribution, drawing through simulation the values used to calculate the index, according to a given law of probability. These simulations are known as "Monte-Carlo simulations", by analogy with draws in games of chance. The number of simulations is chosen so that there is a good approximation of the random distribution of the index, using the survey theory. When the test concerns the localization of values and not the values themselves, simulation involves the permutation of the observed values in the set of possible localizations.

Main statistical tests for the comparison of two groups:

– Chi-square test, Fisher's exact test, for the comparison of frequencies (qualitative variables).

– Z-test, Student's test, Mann–Whitney test, Wilcoxon signed-rank test for the comparison of means (quantitative variables).

– F-tests for the comparison of variances.

– Hotelling's tests for the multivariate comparison of two groups.

Main statistical tests for the significance of an exposure:

– Mantel–Haenszel X^2 test for the significance of relative risk and of the odds ratio adjusted on a confounding factor.

– Breslow–Day and Simon's test for the significance of a ratio.

Box 2.17. *The main statistical tests in epidemiology*

It is quite common to conduct several different tests on the same set of objects, for example to test the implication of several risk factors. Then, each test has its own H0 and HA hypotheses. However, the results are often given an overall interpretation, in the sense of an overall hypothesis H0, which is refuted as soon as one of the elementary tests refutes its H0. This leads us to the multitesting problem [MIL 96]: the more tests we conduct, the higher the probability that one of the tests refutes its H0 by type I error (which is theoretically equal to $1 - (1 - \alpha)^n$, where n is the number of tests, α is the risk of error of each test, if the tests are independent – which is seldom verified). To avoid having a too high risk α' at the level of the overall test, the risk α of each elementary test must be reduced (in order to maintain $1 - (1 - \alpha)^n$ within a confidence interval corresponding to the risk chosen for the overall test), or else it is necessary to modify hypotheses H0 and HA for the overall test (for example, H0 is refuted if at least two elementary tests reject their H0).

Most statistical tools used in spatial analysis rely on the principle of statistical tests.

The multitesting problem often arises in spatial analysis: tests are often conducted in very large numbers (for example, at each point in space), but only an overall conclusion is retained (if there is a point that rejects its H0, then an overall H0 is rejected).

2.4.3. *Methods for model adjustment*

It is common practice to try to model the distribution of observed data by a mathematical function of given form, whose parameters must be determined. The adjustment methods presented here allow the selection of these parameters for the best replication of the observed data by the model.

2.4.3.1. *Maximum likelihood*

According to this approach, what is observed has the highest probability of occurring. The underlying philosophy of this approach is therefore in contrast with that of the statistical test, which assumes that the value observed through datasets may be far from the actual unknown value (assumed to exist) because of the variability due to chance or measurement accuracy. It is related to the notions of causality and determinism (if there is no causality relationship between the sought-for value and the phenomenon assumed to be involved in the determination of the dataset, the principle does not operate. Admitting the principle of maximum likelihood therefore relies on the hypothesis of the determinism of the sought-for value with respect to the studied phenomenon).

In order to distinguish these two approaches, a semantic distinction must be made between probability and likelihood. Probability is used in relation to a parameter, while likelihood relates to a dataset. The likelihood of a situation or a sample is the probability of observing this situation in a set of possible situations, which requires knowledge of a probability law that can be used to determine this set. For example, in the case of a numeric variable, its random distribution is often modeled by a normal law, which is entirely defined by two parameters (mean and variance). In the case of a number of occurrences or of a frequency, a normal law can be used to model the random distribution or Poisson law (also defined by two parameters, mean and intensity) in the case of small counts.

When a hypothesis is formulated on the model pattern and observed data are available, model parameters can be estimated so that it is the best modeling of the observed values. An index is used for this purpose, which is a function of model parameters, allowing the measurement of the matching between the model and the observed data. This index is called the **likelihood** of the model if it corresponds to the probability of generating with the model the obs values observed in reality:

$$\mathcal{L} = P(obs|\theta, \mathcal{M})$$

where $\mathcal{L}$ is the likelihood, P is the probability and θ is the required parameter of the model $\mathcal{M}$.

When the model corresponds to a random variable X with density function f, the likelihood of the parameter θ, given n observations $x_1, \dots, x_n$ (assumed to be independent, hence the product), is:

$$\mathcal{L}(x_1, \dots, x_n |\, \theta) = \prod_{i=1}^{n} f(x_i;\, \theta)$$

Maximizing this likelihood corresponds to the adjustment of parameters θ of the model depending on the observed data, arbitrarily considering that what has been observed is the situation that the model had the highest chance of generating:

$$\theta^* = argmax_\theta\, P(obs|\theta, \mathcal{M})$$

In order to maximize this likelihood, a common practice is to consider its logarithm (which allows the passage from product to sum), and to solve the equations resulting from setting to zero the partial derivatives of likelihood with respect to the variables composing θ.

EXAMPLE.– Let us consider a Bernoulli variable (discrete distribution of probability with two values, and which takes the value 1 with probability p and the value 0 with probability $1-p$). The density function is equal to:

$$P(X=x)=p^x(1-p)^{1-x}, x \in \{0,1\}$$

What is the maximum likelihood estimation of probability p with which a coin flip yields tails? As previously, there have been k tails by flipping this coin n times (observed data). The likelihood of this dataset is given by:

$$\mathcal{L}=p^k(1-p)^{n-k}$$

where $\mathcal{L}$ is expressed as the product of n density functions of Bernoulli variables corresponding to each draw (or, which is approximately equivalent, like the likelihood of the binomial model of parameters n and p).

In order to find the value of p that maximizes $\mathcal{L}$, the derivative of $\mathcal{L}$ with respect to p is set to 0, which leads to $p=k/n$, observed frequency, as in the previous calculation with the expectation.

2.4.3.2. *Bayesian inference*

The principle of maximum likelihood stems from a Bayesian approach.

According to the frequentist approach of probabilities, it is, in principle, assumed that the exact value of the required parameter exists, the events can be replicated and they are subjected to random variability. They are finally analyzed to assess this variability and deduce the value of the required parameter or the interval to which it belongs. In the Bayesian approach, an uncertain hypothesis is formulated, and using an initial idea on the phenomenon (an *a priori* model), the uncertainty of the result is adjusted as events occur (using the calculation of probabilities, and particularly the calculation of conditional probabilities), based on the principle that what occurs is the most likely. The initial model can itself be revised during the process.

There is a philosophical difference between these two approaches: according to the frequentist approach, the exact value of the required parameter is assumed to exist (calling for the choice axiom), but the value calculated from the observed data is subjected to random variability, particularly due to the influence of other parameters on the observed data. Therefore, the uncertainty of the result stems only from the variability of the observed data. In the Bayesian approach, the observed data are considered free from random variability (there is actually only one realization, which is the one observed in reality), the principle of maximum likelihood is enacted and the uncertainty of the result depends on the uncertainty of the event itself (particularly due to other causal factors) and not on the observed data.

The Bayesian approach requires the inclusion in the analysis of an *a priori* model and the gradual use of the observed data, in order to refine the model in real time. Therefore, it allows the use of *a priori* knowledge, which sometimes renders it more efficient. Bayesian inference methods require conditional probabilities and rely on Bayes' theorem.

> Let us consider an event C and an event E. Bayes' theorem stipulates that the probability of occurrence of the event C, knowing that E has occurred, is equal to the probability of occurrence of the event E, knowing that C has occurred, multiplied by the probability of occurrence of C, divided by the probability of E.
>
> $$P(C|E) = \frac{P(E|C).P(C)}{P(E)}$$

Box 2.18. *Bayes' theorem*

Bayes' theorem implies that the knowledge on the probability of the effect E allows feedback to the probability of the cause C and its improvement. The sought-after probability of the cause C is no longer calculated or evaluated using the formal calculation on the observed events C, but on the events E that are considered to be the effects of C. An *a priori* hypothesis is formulated on the distribution of probabilities of C, and then this distribution is modified depending on the events E observed and during the observations.

In an **inferential framework** (the goal is to determine the probability of C depending on the observed data E), Bayes' formula can be written as:

$$P(\theta\,|obs) = \frac{P(obs|\theta).P(\theta)}{P(obs)}$$

Similar to the maximum likelihood, the goal is to estimate θ depending on the observed data. $P(\theta|obs)$ designates the probability of the parameter θ, knowing that obs has occurred, or *a posteriori* probability. The numerator is the likelihood $P(obs|\theta)$, multiplied by the probability $P(\theta)$, which represents *a priori* the probability of the parameter θ (which is known as an *a priori* probability of θ).

Bayesian inference involves replacing $P(\theta)$ with $P(\theta\,|obs)$, the event obs having brought additional knowledge that allows the modification of $P(\theta)$. This knowledge depends on $P(obs|\theta)$ (likelihood of the model), and indicates the degree of causality of obs due to θ. Therefore, the Bayesian approach involves the estimation of the distribution of probability of θ depending on the observed data, and not directly on the "best value".

As will be seen in the remainder of this book, Bayesian inference can be used in spatial analysis to improve the assessment of risks (considered as probabilities) from the observed spatial situations.

EXAMPLE.– Let us consider the same coin-toss game: the Bayesian approach is different from the previous approaches. An *a priori* assumption is made that the distribution of k (number of tails among the n draws) follows a normal law (which approaches the binomial law as soon as $n > 30$). This distribution is given an *a priori* variance (large or narrow, depending on the assumed uncertainty and the risk of error that is taken), and the assumption is made at the start that the coin is not biased (therefore, the mean p of the distribution is *a priori* fixed at 0.5). The observation of results allows the adjustment of the variance of p according to the draws, and the observation of whether the value of p gets significantly farther from 0.5.

2.4.3.3. *Least squares*

Within the framework of the maximum likelihood principle, the least squares method is another way to solve the equation $\theta^* = argmax_\theta\, P(obs|\theta, \mathcal{M})$.

This is an optimization method that involves the adjustment of parameters θ of the model $\mathcal{M}$ – whose form has been *a priori* chosen – by minimizing the sum of the squares of differences between the model and the observed data:

$$RSS = \sum_i (x_i - \mathcal{M}(i))^2$$

EXAMPLE.– If a simple linear model $\mathcal{M}(x) = ax + b$ is chosen, the line determined by the least squares method corresponds to the regression line. The coefficients a and b result from solving so-called "normal" equations:

$$a = (n\sum_{i=1}^{n} x_i y_i - \sum_{i=1}^{n} x_i \sum_{i=1}^{n} y_i)/(n\sum_{i=1}^{n} x_i{}^2 - \sum_{i=1}^{n} x_i \sum_{i=1}^{n} y_i)$$

$$b = \frac{1}{n}(\sum_{i=1}^{n} y_i - a\sum_{i=1}^{n} x_i)$$

The model can also be assessed by minimizing the likelihood function of the model.

If an estimation of the difference in type σi that affects the measurement of x_i is available, the influence of the measurement i in the sum RSS can be weighted and the quantity X^2 can be minimized, rather than S:

$$X^2 = \sum_i (\frac{x_i - \mathcal{M}(i)}{\sigma i})^2$$

When the quality of the model is assessed on observed data (which have not been used in determining the parameters of the model), the mean squared error can be used:

$$E_{RMS} = \sqrt{\frac{\sum_{i=1}^{N}(x_i - \mathcal{M}(i))^2}{N}}$$

where N is the number of observations x used for the evaluation.

2.4.3.4. *Akaike information criterion (AIC)*

Akaike information criterion (AIC) is an estimator of the quality of a statistical model. It aims to minimize the number of parameters of a model. In fact, the likelihood of a model can always be increased by adding parameters. AIC allows the quantification and therefore optimization of the relationship between the number of parameters and the maximum of the likelihood function of a model. It thus allows a model to be chosen among several others, choosing between model simplicity and adjustment quality:

$$AIC = 2k - 2ln(L)$$

where k is the number of parameters to be assessed and L is the maximum of the likelihood function of the model. For example, for a least squares-based assessment of a model with N observations whose errors follow a normal distribution, the following formula can be used:

$$AIC = 2k + N(log\left(\frac{RSS}{N}\right))$$

to compare several models.

The calculation of AIC is provided by most of the statistical software.

2.4.4. Several distributions and models

Knowledge of several models and distributions of probability, which are very commonly used in statistics, is very useful, particularly in order to model random variability and understand the choice of *a priori* models in Bayesian inference: binomial, normal, Poisson, chi-square, gamma, etc. These distributions correspond to various types of variables that will have to be considered in the remainder of this book: count data, corresponding events (cases, mosquito counts, etc.), frequencies (ratios between count data), numeric variables corresponding to measurements (size, surface temperature, water quality, etc.). They have strong interrelationships, which are briefly described below.

2.4.4.1. *Bernoulli variables and the binomial law*

A Bernoulli variable is a random variable that can take two values, with the probabilities p and $q = 1 - p$, respectively. A Bernoulli variable follows a Bernoulli law of parameter p. A Bernoulli variable allows the formalization of a dichotomous (success/failure) event:

$$P(X = 1) = p \; et \; P(X = 0) = (1 - p)$$

Expectation is equal to p, and variance is equal to $p(1 - p)$.

By definition, the binomial law of parameters $n \in \mathbb{N}^*$ and $p \in [0,1]$ is the law of the sum of n independent Bernoulli random variables of the same parameter p. The binomial law thus represents the sum of successes in a series of individual independent Bernoulli trials of the same probability p. It is a law of discrete probability (the random variable X takes only integer values), whose value is given by the formal calculation of probabilities:

$$P(X = k) = \binom{n}{k} p^k (1 - p)^{n-k} \text{ for } k = 0, \ldots, n$$

Expectation is equal to np, and variance is equal to $np(1 - p)$. The distribution function ($F(x) = P(X \leq x)$) can be easily expressed; however, the calculation is not easily done by a computer due to the binomial coefficients and factorials that can rapidly take very high values. Therefore, the binomial law is often approximated by the Poisson law or the normal law.

Another distribution is the negative binomial law: it expresses the distribution of probabilities of the random variable X corresponding to the number of failures required before obtaining n successes in a series of Bernoulli trials (independent and of the same probability p):

$$P(X = k) = \binom{k+n-1}{k} p^n (1 - p)^k \text{ for } k = 01,2, \ldots$$

Its expectation is equal to $\frac{n(1-p)}{p}$, and its variance is equal to $\frac{n(1-p)}{p^2}$. It is used for modeling (for example, to calculate the probable number of mosquito bites before an infected bite occurs).

2.4.4.2. *Poisson distribution*

The Poisson law is a law of discrete probability that yields the probability of observing a number k of events depending on the mean frequency λ of each event per unit (of time, surface, volume, etc.). The Poisson law also allows the calculation of the probability of observing a number of events in a unit of time or on a surface by chance. The events are assumed to be independent with respect to the unit. The Poisson law is expressed as a function of two parameters, k and λ:

$$P(X = k) = e^{-\lambda}\frac{\lambda^k}{k!} \; with\, \lambda > 0$$

The expectation and the variance of X are both equal to λ.

The binomial law (of the parameters n and p) can be approximated by the Poisson law (of the parameter $\lambda = np$) when the number of trials n is large (in practice, above 50) and the expectation λ is low (in practice, below 5). Thus, the Poisson law is used instead of the binomial law in order to calculate the probability of rare events.

2.4.4.3. *Normal distribution*

Let us consider a random continuous variable X (which can take real numeric values) with a density function. The probability $P(a < X < b)$ of such a variable is expressed as a function of its density of probability f:

$$P(a < X < b) = \int_a^b f(x)dx$$

which corresponds to the area under the curve of function f between a and b.

The density of probability of the normal law $N(\mu, \sigma)$ is:

$$f(x) = \frac{1}{\sigma\sqrt{2\pi}}e^{-\frac{1}{2}(\frac{x-\mu}{\sigma})^2}$$

The expectation of $N(\mu, \sigma)$ is μ and its variance is σ^2. When the variable x is already reduced-centered ($\mu = 0, \quad \sigma = 1$), the normal law is also called reduced-centered. The plot of f is the famous bell curve or the Gauss curve (the normal law is also called the Gauss distribution).

The normal distribution is particularly important in statistics: the central limit theorem shows that the distribution of the sum of independent random variables with comparable variances follows a normal law. The normal law is thus omnipresent in nature, as it models phenomena whose random distribution is due to the addition of a large number of small disturbances, whatever the distributions of these disturbances.

The normal law can also be used as an approximation to the binomial law. Let us consider a random variable X_n that follows a binomial law with parameters n and p, and the reduced-centered variable $Z_n = \frac{X_n - np}{\sqrt{np(1-p)}}$ corresponding to X_n. The de Moivre–Laplace theorem indicates that $lim_{n \to \infty}(P(a \leq Z_n \leq b)) = \int_a^b \frac{1}{\sqrt{2\pi}} e^{-\frac{x^2}{2}} dx$. Therefore, the binomial law of parameters n and p can be approximated by a normal law of mean np and variance $np(1-p)$. In practice, when $n \geq 30, np \geq 5$ and $n(1-p) \geq 5$, the differences in probabilities are very small. These application conditions are therefore often found in the literature.

The Poisson law with intensity λ can also be approximated by a normal law of mean and variance λ. In practice, the differences in probabilities are very small as soon as $\lambda > 25$.

Spatial analysis uses several calculations that require the product X of random variables X_i, and not their sum. The central limit theorem cannot be directly applied to X, but as the logarithm of product X is equal to the sum of the logarithms of random variables X_i, this sum tends to a normal law if $log(X_i)$ and hence X_i satisfy the conditions for applying the central limit theorem. The product X is then said to follow a log-normal law.

2.4.4.4. *Gamma distribution and chi-squared law (χ^2)*

The importance of the chi-squared law (χ^2) lies in the role it plays in statistical tests, particularly in the tests for the comparison of counts and in the tests for comparing a theoretical law to an observed distribution. It is used to study the sum of squares of random variables, which is often the case, since due to mathematical properties, it is preferable to study the squares of differences than the absolute values of differences.

χ^2 designates the random variable that is the sum of n reduced-centered normal random variables Ti that are independent:

$$\chi^2 = T_1 + T_2 + \cdots + T_n$$

This variable ranges between 0 and $+\infty$, whose density of probability is:

$$f(\chi^2) = \frac{1}{2^{\frac{n}{2}}\Gamma(\frac{n}{2})}\chi^2 e^{\frac{\chi^2}{2}}$$

It is said to follow the law of χ^2 with n degrees of freedom. Γ is the gamma function, which is the continuous extension of the factorial function. The law of χ^2 can be assimilated to the normal law when the number of degrees of freedom is above 30.

For example, let X be a random variable following a law of probability P. The values of X are classified into k modalities. For a class, the reduced-centered gap between the theoretical and the observed counts can be considered a reduced-centered random variable as soon as the binomial law can be approximated by the normal law. The sum of squared gaps for all the classes provides a measure of the distance between the observed distribution and the theoretical distribution. It is a random variable that follows a distribution of χ^2 with $k-1$ degrees of freedom, and therefore depends only on the number of classes and not on the P law. The classical χ^2 test is a direct application of this result.

SUMMARY.–

– From an initial spatial distribution of individuals, the spatial distribution of those affected by a health phenomenon is the result of many processes involving the relationships between agents (hosts, vectors where appropriate) and the relationships with environmental parameters (mobile or immobile).

– The initial spatial distribution of agents must be known, but it is not among the processes to be analyzed.

– Contagion processes and processes of exposure to risk factors are the main elements to be identified, based on the spatial distribution of hosts.

– The objects that are subject to analysis must be clearly specified, as well as their domain of definition and their spatial implantation.

– In epidemiology, spatial analysis is complementary to classical statistical analysis.

– Overall analyses provide information on the characteristics of the overall spatial distribution (homogeneity, overdispersion, autocorrelation, tendency, centrality, pattern), while local analyses provide information on the local variations of spatial distribution (excessive risk, clusters).

– Spatial–temporal modeling allows the simulation and analysis of processes of emergence, diffusion and extinction.

3

Spatial Data in Health

This chapter presents the type of data used in epidemiology and health geography. It presents in particular the specificities associated with health data localization.

3.1. Introduction

The main sources of data on the factors that play a determining role in health come from public or private health bodies (hospitals, networks of pharmacies, Health Department, health insurance departments). These institutions provide data on the healthcare system, on the use of care and on the consumption of healthcare and medication, sometimes at the individual level, but most often at the aggregated level. The quality of health data collection and management varies from one country to another and often, particularly in developing countries, from one region to another within the country: densely populated urban areas are often significantly better covered than rural areas, where the population density is lower. "Data blindness" is quite often noted: the most vulnerable regions for health have also a lower quality in reporting epidemiological data.

Specific surveys are conducted when there is a need for additional information on the presence of a pathogen, a disease, a prevalence, or the relationship between a health variable and an assumed risk factor. Observations, surveys, collections, captures and field measurements are conducted in order to characterize the presence or measure the abundance of an organism, vector, reservoir, etc. These studies often involve surveys, and do not collect exhaustive data. Some physicians are also involved in networks for epidemiological passive surveillance and real-time collection of epidemiological data such as the Sentinelles program in France, which allows the follow-up of the space-time evolution of some infectious diseases.

Environmental data (natural, urban, technological) and demographic data are essential for most studies concerning risk. The interpretation of satellite images or aerial photography is also an important source of data (land use, temperature, humidity, state and type of vegetation, etc.).

3.2. Health data

Strictly speaking, epidemiological data refer to data collected by the healthcare system or through specific surveys on the health state or health events for a given population. In the systemic view presented in Chapter 1, they essentially concern individuals. Other data have been used in epidemiological or geographical analyses. They concern the description of the healthcare system, the health system, the infrastructures and transportation networks, flows, healthcare pathways, social environment, natural environment, etc.

3.2.1. *Various types of data for individuals*

For individuals, we can distinguish between:

– a state: the state of an individual for a health-related factor at a given moment. For example: individual data (age, gender, size, etc.), genetic data, immune status, exposure to a risk factor, etc. Clinical studies collect exposure data or therapeutic effects from cohorts of individuals, at various dates and for various products;

– an event: the occurrence of a change in the health of an individual. For example: infection, change in the immune status, disease expression, hospital admission, medical care, etc. Very often, an event includes a variable providing the date, sometimes the place.

The objects (state or event) are characterized by attributes. As already seen in the previous chapter, a classical approach involves the following types of attribute:

– qualitative values (name, state of health, name or code of disease, social security number, etc.);

– Boolean values (yes/no, sick/non-sick, etc.);

– counts (integer values);

– relative numeric values: quotient, ratio, rate (incidence, prevalence, densities, odds ratio, relative risk, etc.);

– numeric values that represent quantities or measurements (age, weight, cost, distance, duration, temperature, etc.);

– dates.

3.2.2. *Individual and aggregated health data*

Epidemiological data may describe individuals or groups of individuals:

– the data corresponds to either cases (individuals considered positive as result of a test or diagnosis) or (individual) events, with attributes describing the individual's general characteristics (age, gender, behaviors, etc.) or the characteristics of his state (date of the event, symptoms, severity, etc.). Data directly collected from healthcare systems or health insurance systems belong to this category. Data can be exhaustive (collection of all the cases) or obtained through a survey (on a population sample). Verifying hypotheses on the studied phenomenon generally requires information on non-cases: from a census on all the individuals, from the sampling frame or from the overall sample for a survey, or by setting up another sample for non-cases (control sample). Non-cases must be described by the same general attributes as the cases. Data collection is often limited to a place (country, region, city, etc.) and duration (between an initial date and a final date);

– alternatively, the data describing individuals are aggregated (by those who produce them) in order to describe other more general objects, using sums, means, rates, frequencies, or even statistical distributions. In this case, individual data are no longer accessible. It is very common to have data that correspond to counts of cases aggregated according to a criterion (spatial unit, age group, gender, etc.). An aggregate based on belonging to a space (region, department, village, etc.) is very often used by data producers in order to summarize the health status in a population on a territory. When data correspond to counts of cases, it is important that overall counts of individuals are available in order to calculate rates;

– secondary data may correspond to more complex objects, such as care pathways. These objects are built from other objects, provided that attributes can be joined to allow their construction.

Available data (that can be accessed) must be differentiated from the results of analyses, particularly statistical analyses, which allow the aggregation of results from available attributes (aggregation into several groups uses values of a qualitative attribute, spatial or not).

3.2.3. *Description of the healthcare system*

The healthcare system (care, prevention, pharmacy, well-being) is an essential element in any healthcare study, as it is generally the source of data, particularly of epidemiological data. Health geography is particularly interested in the analysis of the healthcare system and in the analysis of the relationships between healthcare systems and populations (accessibility, inequalities, practice population, construction of health territories, etc.).

The description of the healthcare system rather relies on an individual description of objects than on aggregated data: list of hospitals, healthcare centers, consultation centers, physicians, dentists, pharmacies, medical laboratories, etc. The objects are described by attributes, such as:

– hospitals: number of beds, number of physicians, number of nurses, number of specialties, volume of medical acts, etc.

– physician: name, address, specialty, type of contract, type of practice, etc.

– pharmacy: name, address, turnover, number of pharmacists and number of pharmaceutical assistants.

3.3. Spatialization of epidemiological data

3.3.1. *Localization in space*

By definition, an individual (object or event) is localized if it contains information on its position in space.

Whenever mentioned, epidemiological data refers to data concerning a health event (a change in the health state). Since individuals are not immobile, it is difficult to know the actual position of an event. A selected position of the individual (residence, work place, leisure place, etc.) is generally used, and this does not necessarily correspond to the individual's position at the time or during the event. Position is given either by the geographic coordinates of the place, an address, or by the individual's position in a more general geographical entity (hospital, village, district, etc.) that can be localized, if needed. In order to be used (particularly by GIS), addresses must be able to be converted into geographical coordinates: it is the purpose of geocoding. There are many software programs that allow geocoding (for example, MMQGIS plugin for QGIS, Geocoder in ArcMAP).

Data on individuals' exposure to a risk factor are even more difficult to obtain, as their movements over time should be followed, and exposure should be evaluated or calculated based on these movements (Figure 3.1).

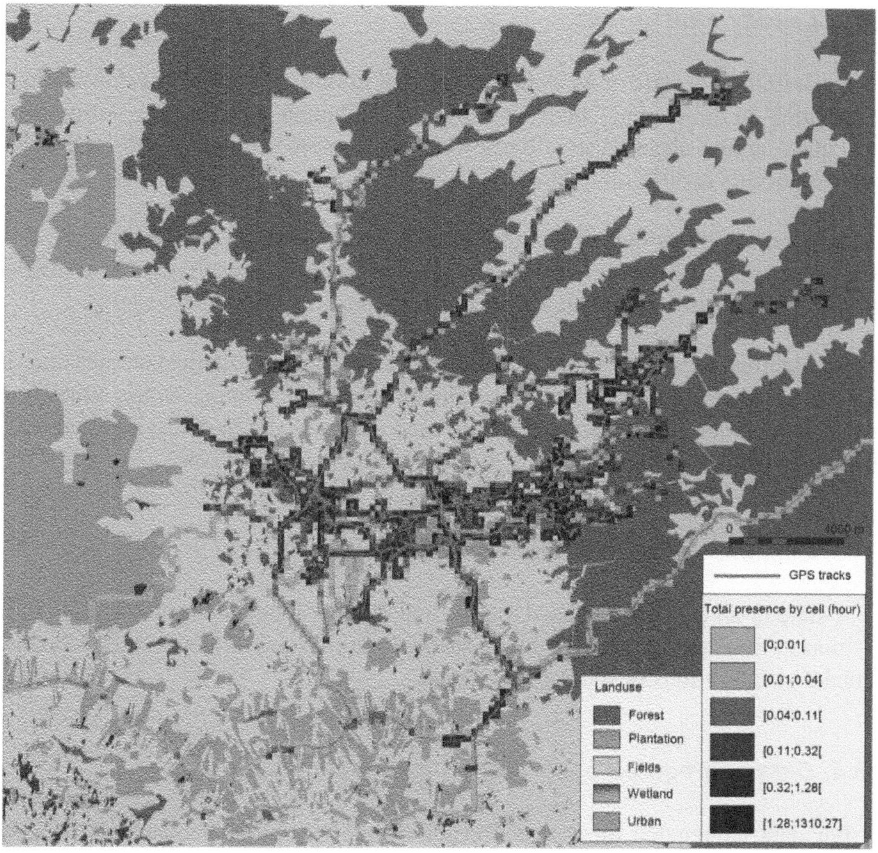

Figure 3.1. *GPS tracking of movements of a cohort allows the calculation for each individual of the time spent in a place and their exposure to various environments. It also allows the characterization of the most crowded passage places. For a color version of this figure, see www.iste.co.uk/souris/epidemiology.zip*

As previously mentioned, individual epidemiological data (the objects of which are individuals) should be distinguished from epidemiological data resulting from an aggregation, and mentioned objects of which are entities used to group together individuals (hospital, district, age group, etc.). Individual data can be spatialized with coordinates (usually a point), or with addresses, or by reference to a geographical division (municipalities, districts, census areas, etc.). This spatialization may be accurate, but there will often be ambiguity between the localization of the individual and the location of the health event that concerns it or is the object of study (infection, exposure, transmission, etc.).

3.3.2. *Localization in time*

By definition, a state or an event is localized in time if it contains information on its position in time: date, hour or duration. Although simple, this definition is difficult to apply; accuracy is often low, as the precise moment when an event or exposure occurs is difficult to know.

Hospital registries contain admission and discharge dates, and sometimes dates of symptoms or of healthcare provision. Date of medical consultation or admission to hospital does not correspond to date of symptoms.

Availability of time information of events enables chronological studies and the analysis of the evolution in time of a phenomenon. These analyses use time series analysis techniques. The most commonly used analyses aim to detect a global trend, an overall tendency, a periodicity, seasonal fluctuations, cycles, etc.

3.3.3. *Localization in time and space*

A state or an event is localized in time and space if it contains an indication of its position in time and space. It is then possible to jointly use this information in order to make visualizations or space-time analyses.

3.3.4. *Data aggregated according to a spatial criterion*

For data aggregated on a spatial criterion, the grouping of objects correspond quite often to immobile geographical units (administrative division, medical entity, etc.) whose localization is well defined in time and space. Since the initial individual data are aggregated, the attributes correspond to count ratios, or to the presence/absence of a characteristic in the aggregation object. The studied object is no longer the individual, but the unit of aggregation. Once again, there is ambiguity between the localization of health events and the localization of geographical units. Unfortunately, spatial relationships between initial objects (other than the fact of belonging to the same spatial unit) are lost. These geographical units generally have other data – for example, geographical or environmental data – that allow the calculation of rates or the estimation of probabilities based on a model. Many problems arise when comparing geographical units used for the aggregation, as the accuracy of these calculations is not necessarily the same for all the units. This is one of the problems raised by statistical cartography, which will be addressed in Chapter 4.

Whenever ethically and administratively possible, it is preferable to have non-aggregated data. GIS have tools for making subsequent aggregations when needed (notably for the calculation of rates or evaluation of probabilities) and for choosing different levels of aggregation.

While various levels of aggregation can be used in statistical analyses, it is difficult to disaggregate data that are already aggregated. For example, to be able to disaggregate, at the level of municipalities, the data aggregated at the district level, strong hypotheses must be formulated on the distribution of values of municipalities in the districts. From a practical perspective, *downscaling* methods, used in particular in meteorology [TRZ 14], or dasymetric methods (using density measures) can be used in order to disaggregate data.

In summary, the level of aggregation at which data are available determines the possibilities of analysis:

– when data are available at the level of individuals, and they are accurately localized (in time and space), all statistical, spatial and spatio-temporal analyses are possible;

– when data are available at the level of individuals, but they are not accurately localized, some spatial analyses cannot be conducted. This is notably the case when localization solely stems from belonging to a space;

– when data are not available at the level of individuals, but only for counts in groups, it is not possible to conduct analyses on individuals themselves. Statistical, spatial and spatio-temporal analyses are only possible on these groups, and not on the individuals that compose them.

3.3.5. *Ethics and localization*

When accurate, the geographical localization of an individual provides information that may violate basic ethical rules, notably by breaching individuals' anonymity. These rules are not the same for all the studies and surveys: localizing medical practices or physicians in order to study the localization of healthcare provision may be accepted by an ethics committee, while localizing patients' residence, which is conducive to breaching their anonymity, will very likely face an opposition.

3.4. Sources of data

3.4.1. *Epidemiological data*

Epidemiological data can be collected from:

– healthcare systems (hospitals, healthcare centers, prophylactic centers, sentinel physicians, veterinarians, etc.). Data describe individuals or healthcare pathways, with attributes on the state of health, pathologies, and general attributes of the individual or their behavior such as age, gender, weight, etc. These attributes can be qualitative (profession, gender), numerical representing quantities (number of children), numerical representing values (weight, voltage, etc.) or ratios (body mass index, etc.). Some diseases are notifiable. There are disease-specific registries (cancer registry, for example). Their objective is an exhaustive collection of disease-related data;

– surveys and representative population samples, which have been chosen to allow the evaluation of the characteristics to be detected in the global population. These surveys are not solely related to patients, particularly in the epidemiological surveys of risk factors.

In France, as in many countries, health data are considered "sensitive" according to Directive 95/46/EC Directive and to the law on "Data processing and civil liberties". The Babusiaux report (2003) recommended the transmission of these data, on condition of anonymity, not only to public health insurance agencies (CPAM – *Caisses Primaires d'Assurance Maladie*), but also to mutual insurance companies.

The two main French health databases are:

– Database of the Program for the Medicalization of Information Systems (*Le Programme de médicalisation des systèmes d'information* – PMSI), which collects information on the hospital stays in France. It is managed by the Technical Agency for Information on Hospital Care (*Agence Technique de l'Information sur l'Hospitalisation* – ATIH). All access to individual data must be authorized by CNIL;

– National System for Inter-regime Information of Health Insurance Agency (*Système National d'Informations Inter-Régimes de l'Assurance Maladie* – SNIIRAM), which has been progressively implemented by the health insurance agency since 2004. It is the French national database dedicated to medical pricing containing information on the consumption of reimbursed healthcare. Many data are now available online on the site www.ameli.fr.

The site www.data.gouv.fr also contains many datasets, some of which are localized.

The sites Eco-Santé (www.ecosante.fr) and Score-Santé (www.scoresante.org) of the National Federation of Health Observatories (*Fédération nationale des observatoires de santé* – FNORS) and the Institute for Research and Documentation in Health Economics (*Institut de recherche et documentation en économie de la santé* – IRDES) provide many health-related data (epidemiological, pharmaceutical, healthcare system, health economy, demography). Data are available at various geographical levels of aggregation, from regions to departments. Numerous epidemiological data can also be found at the site of InVS (invs.santepubliquefrance.fr).

Open Medic database consists of a set of annual databases concerning the medicines delivered through pharmacies. Real-time data on the consumption of medication are collected through some pharmacies by institutions in the private sector (CeltiPharm, IMS Health). CépiDC (INSERM) data on the medical causes of death are also worth mentioning.

Box 3.1. *Access to epidemiological data in France*

The development of *Big Data* and its consequences for medical records will very likely impact the overview of health data. These questions are core issues for certain research teams, such as the team-project "Big data in health" of LTSI-UMR Inserm – University of Rennes 1.

3.4.2. *Geographical and environmental data*

Many other data can be involved in epidemiological studies and health geography studies. Epidemiological data are also jointly used with other data, such as:

– demographic and socio-economic data resulting from population or habitat censuses, agricultural censuses, etc. For ethical and confidentiality reasons, census data are always aggregated and do not describe individuals, but groups;

EXAMPLE.– In France, INSEE provides socio-demographic data aggregated by municipalities or by IRIS (fundamental unit for dissemination of infra-municipal data, containing minimum 2000 residents).

– data describing the environment, such as soil usage, relief, climate, air quality, type of soil, buildings, etc. These data do not describe individuals, but objects, which are often geographical objects localized in space [HER 07];

– various geographical objects, allowing in particular the aggregation of epidemiological data on administrative bases or structures (administrative divisions, pathways, squaring, etc.);

– data on the healthcare system: description and localization of health infrastructures (healthcare centers, hospitals, etc.), census and localization of health personnel (physicians, nurses, midwives, etc.) and private medical practices (medicine, dental, etc.);

– data on the infrastructures related to accessibility or to traffic accidents, such as road networks and transportation systems.

Veterinary surveillance data are also important for the study of zoonoses (diseases transmissible from animals to humans). There are sometimes data on domestic animals (for example, files on dogs and cats in France, and files on horses at the European level). Administrative services in charge of farming management may have accurate information on the characteristics and localization of farming places. Sentinel veterinarians allow the monitoring of infectious pathologies, particularly in equine pathology, in a similar way to the networks for monitoring some human infectious diseases. Agricultural censuses also contain numerous information on farming and farm animals.

Information on the localization of wild animals has also been collected, especially during research operations. For example, there are tracking devices for wild birds and for migratory birds (particularly for the study of flu and West Nile virus), RFID chips implanted in reintroduced wild animals with the purpose of recording the passages of animals at some points of the territory, etc.

3.4.3. *Access to geographical data*

There are numerous geographical data available, but the offer is disparate. Internet access has significantly increased in recent years. Many sites provide Internet users with environmental data, for example:

– global climate data can be accessed on the site www.wordclim.org, with a geographical resolution of 1 km^2, and a time resolution of one month (monthly averages of minimum and maximum temperatures, averages of rainfall, sunshine, wind speed, water vapor pressure, for the period between 1970 and 2000). Lower resolution (approximately 250 km^2) data are available for the period between 1901 and 2006 on the site www.cgiar-csi.org/data (daily average temperature, monthly minimum, monthly maximum, rainfall, etc.);

– topographic data with a 90-m horizontal resolution are available on the site www.cgiar-csi.org/data/srtm-90m-digital-elevation-database-v4-1. A 30m resolution is also available in several regions of the world;

– in 1985, the European Union has established high accuracy land cover observation services for European countries (based on satellite images with a 20m resolution). Data are available on the site: www.land.copernicus.eu. Global data on land cover (GlobCover, 2010) at a lower resolution (300 m per pixel) are also available and can be downloaded from the site of the European Space Agency (http://due.esrin.esa.int/page_globcover.php). Further environmental data are available on the ESA site;

– Copernicus is the Earth observation of the European Union, and provides access to many images taken by SENTINEL 1 and SENTINEL 2 satellites;

– OpenStreetMap site provides cartographical data based on GPS coordinates captured and freely shared. OpenStreetMap data can also be downloaded in tiles from the site, available at: https://openmaptiles.com;

– similar to Google Map, DigitalGlobe provides access to a worldwide coverage of high resolution images.

On the contrary, many data are not readily available, for example the accurate localization of health practitioners in the private sector and evaluations of their patients.

While most developed countries have implemented structures allowing Internet access to health data (epidemiology and public healthcare systems), administrative data (administrative divisions, statistical data), demographic data (population censuses and indicators), infrastructure data and territorial planning (agricultural censuses, transportation networks, etc.), the situation is not the same in many developing countries.

SUMMARY.–

– The main sources on the factors determining the state of health or the state of the healthcare system originate with public or private health institutions: hospitals, networks of pharmacies, health departments, health insurance agencies and sentinel physicians.

– Epidemiological data are generally aggregated according to their belonging to a geographical division, and are not available at the level of individuals.

– Environmental and demographic data are essential for most studies concerning risk.

– Accurate localization of health events is difficult to obtain. Localization is most often given by belonging to a geographical entity.

– There is often ambiguity between localization of the individual and localization of the health event that concerns the individual and is the focus of study (infection, exposure, transmission, etc.).

4

Cartographic Representations and Synthesis Tools

4.1. Introduction

4.1.1. *Why use mapping methods?*

Cartography is used to provide local information (on what there is or what happens in a specific place) or to represent and visualize a global situation and global or local spatial relationships between the values of localized objects.

With cartography, it is possible to rapidly apprehend a spatial situation in its entirety. The human brain has strong analytical abilities for visual interpretation. It has the natural instinct of very rapidly and spontaneously detecting overall tendencies (gradients, directions, centrality) and specific local situations, such as clusters or contrasts, which are considered as abnormal situations compared to a "normal" situation that would be more or less uniform.

Cartography applied to epidemiology and health geography mirrors these visual analytical abilities. It allows the visual detection of overall tendencies with little ambiguity. Visual analysis detects overall tendencies that are practically always global events whose probability of occurring by chance is very low. This is because of dimension 2, which squares the number of possible situations. The majority of these possible situations are spatial situations with no specific geometric characteristics, which the brain interprets as more or less uniform. In this context, the human eye easily detects any situation with an overall specific geometric characteristic, and notably a particular pattern (anisotropy, centrality, gradient, periodicity, clustering, etc.), as its occurrence, among all possible situations, is very unlikely.

Unfortunately, and especially when no global pattern can be detected, the brain tries to interpret individual values and local situations on this image at the same time, by taking them out of their overall context. It is much less efficient in doing this, as it does not have the capacity to analyze the actual probability of these situations or local values. Therefore, the visual interpretation of the image can become subjective and risky. The eye intuitively focuses on what it perceives as abnormal, and so it searches for abnormal local situations. For example, the brain detects and interprets as abnormal individual values of an area or the position of an object when they differ from their neighbors, as if continuity was an implicit rule in nature. It always interprets local clusters as unlikely situations, despite clusters often occurring by chance in the global probability distribution.

It is very difficult to comprehend the randomness of a local spatial situation. The human brain follows the principle of maximum likelihood, assuming *a priori* the existence of cause and effect relationships between place and event, which could justify the occurrence of the event. It finds it very difficult to anticipate random variability and, in particular, to consider that a spatial distribution could be "abnormal" solely by chance. In fact, it has great difficulty in considering that the occurrence of an event is, to a large extent or entirely, independent of its location.

Because of this limitation of the human brain, precaution is recommended when using visual analysis. Visual analysis should only be used in the development of models or hypotheses that are accepted or refuted either through statistical and spatial analysis methods, which will be presented in the next chapters, or through inductive reasoning.

4.1.2. *How to use mapping?*

Localized health data can be events or the aggregation of events into geographical units (counts, mean values or ratios). The location of events is often represented by a point.

One–off events can therefore be represented at their respective locations simply by a unique symbol – to address the question "what is it or what happens in a specific place" – or by a symbol whose graphical characteristics depend on the value of the events' particularities – to illustrate a thematic issue. Several events can be located in the same place, which is not easy to represent. These events can also travel in time. To overcome these challenges, events are generally aggregated into immobile geographical units (points, lines or areas). It is these units which are marked on a map, rather than the events themselves. Then, we represent the value resulting from the aggregation of events (counts, mean values, ratios) or from a more complex calculation of the events before aggregation (value yielded by a statistical calculation, a geometric calculation or by a model). Alternatively, we can use methods to estimate an arbitrary point in space by aggregation of the events occurring in the vicinity of that point. In all the cases, mapping is the final stage of the data analysis and its graphical representation.

In addition to the representation of events or aggregated and modeled values, mapping can also be used to represent elements that synthetically summarize a parameter of the spatial distribution. For example, we will show how to calculate and represent the information on the centrality, spatial dispersion and direction of a cloud of point objects. We will also present how to represent a spatial tendency using an interpolation to fill all the space from a discrete domain.

Mapping involves many rules that are part of the terminology of "semiology of graphics" [BER 67]. Various "visual variables" can be used to represent data: shape, orientation, color, value, grain and size. A graphic solution is efficient when the data properties correspond to the properties of the visual variable representing them.

GIS are now the main tools used to create thematic maps. They are sometimes associated with drawing software programs or with specialized software programs for cartographic editing. The principle of automated cartography is relatively simple; it uses the classical cartographic language (made of figures, patterns and colors) and associates parameters of this language (figure, size, color, grain, thickness, etc.) with descriptive values of the objects to be drawn [BEG 00]. The general principles of automated cartography are provided in the Appendix (available online at: www.iste.co.uk/souris/epidemiology.zip), and examples can be found throughout this chapter.

SavGIS	The application of semiology of graphics uses the cartographic browser that can be launched using the icon in the menu bar or by a right click in a frame.
ArcGIS	Data visualization uses ArcMAP module.
QGIS	QGIS visualization involves the same visualization principles as ArcGIS.

Table 4.1. *In practice*

Maps are sometimes grouped in an atlas. With interactive atlases, the user can create maps by choosing certain parameters, based on either the objects or values to be represented, or on the visual parameters.

In France, the regional health agencies (*Agences régionales de santé* – ARS) in collaboration with the Technical Agency for Information on Hospital Care (*Agence technique de l'information sur l'hospitalisation* (ATIH)) provide interactive maps intended for healthcare professionals, with detailed figures on the healthcare provision and consumption by region, department and also by canton and municipality. Nevertheless, these tools are not connected to any other data. They do not include access to healthcare, outside the hospital activities, nor do they include the socio-economic characteristics of these areas, which are useful for analyzing health-determining factors.

Box 4.1. *Interactive atlases of health*

Spatio-temporal data are very difficult to map. Animated videos, which can only be displayed on a screen, can represent the evolution of data in time and space.

4.2. Cartographic representations

4.2.1. *Mapping events or health status*

To map events (object localization, occurrence of a case, of a change of state, etc.), we use a one-point representation that assigns a symbol to each event. It is a delicate task when two events are located at the same place, as the two symbols overlap. The map of events may be difficult to read if there are many events and their symbols overlap. A way to avoid these difficulties is to represent the number of events after aggregation in a geographical unit rather than representing the events themselves (Figure 4.1). The choice of the aggregation unit is obviously key; it must ensure an accurate map reading that answers a particular question (for example, "what is there at a specific place"), while remaining graphically clear (by limiting the number of aggregation units). In this section, we will see how interpolation techniques can be used to avoid aggregation-based methods and geographical division.

Mapping of events, either individual or aggregated, is a raw image of the geographical features of the phenomenon. This image does not consider the potential probabilities of the occurrence of an event depending on place.

4.2.2. *Mapping rates: prevalence, incidence, risk and odds ratio*

Many rates are used in epidemiology: prevalence, incidence, relative risk, odds ratio, attributable risk, etc. A rate always corresponds to a count of data or events aggregated within a group. In epidemiology and health geography, it is very common to use geographical units to calculate counts and deduce rates in these units by using demographic or environmental data.

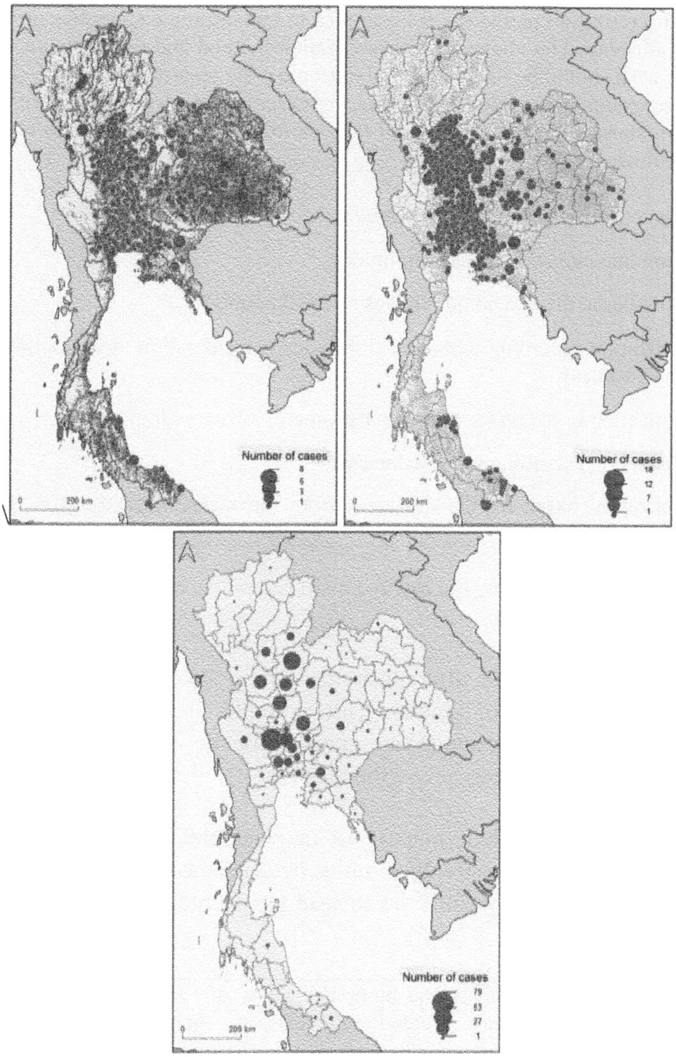

Figure 4.1. *Aggregation of avian influenza cases in Thailand, in various geographical units, from data on villages (basic unit of the localization of cases): sub-districts, districts and provinces*

A discretization is required in order to represent the rate values with colors applied to areas or inside proportional symbols (Figure 4.2). Many discretization methods are available: the choice of method depends on the distribution of values and on the focus of the study [RIC 98].

Many discretization methods can be applied to transform a quantitative attribute into a qualitative attribute. The choice of a discretization method and of the number of classes depends on what should be illustrated, and must meet the requirements of the semiology of graphics (for example, the number of classes represented on a map should not exceed 10). First and foremost, the characteristics of the distribution of the variable to be discretized should be known.

The most commonly used methods are:

– division into classes of equal amplitudes;

– quantile-based discretization (classes of equal counts);

– discretization according to natural thresholds (consideration of discontinuities in the distribution of values);

– discretization based on (arithmetic or geometric) progression;

– discretization according to nested means;

– discretization based on the deviation from the mean, depending on the standard deviation.

Box 4.2. *Discretization-based classification methods*

The color gradient used after discretization is also important: we can use a lightest to the darkest color gradient (for example, from yellow to red), or two successive gradients (for example, one gradient for the values below the mean, and another gradient for the values above the mean).

Colored areas (choropleth maps) are in principle not recommended for the representation of rates in geographical units, because the area of the units influences the reading. Proportional symbols are instead preferable. For further details, please refer to section 4.2.4.

A choropleth map associates the numerical value of each geographical unit with a color. Numerical values are divided into classes, each geographical unit being represented by the color corresponding to its class. Colors are chosen on a scale of graduated shades. The map thus allows the visual comparison of the values of various geographical units (Figure 4.2). This visual comparison can be falsified by several parameters, such as area differences or differences in the significance of ratios, or the range of colors used.

Box 4.3. *Choropleth maps*

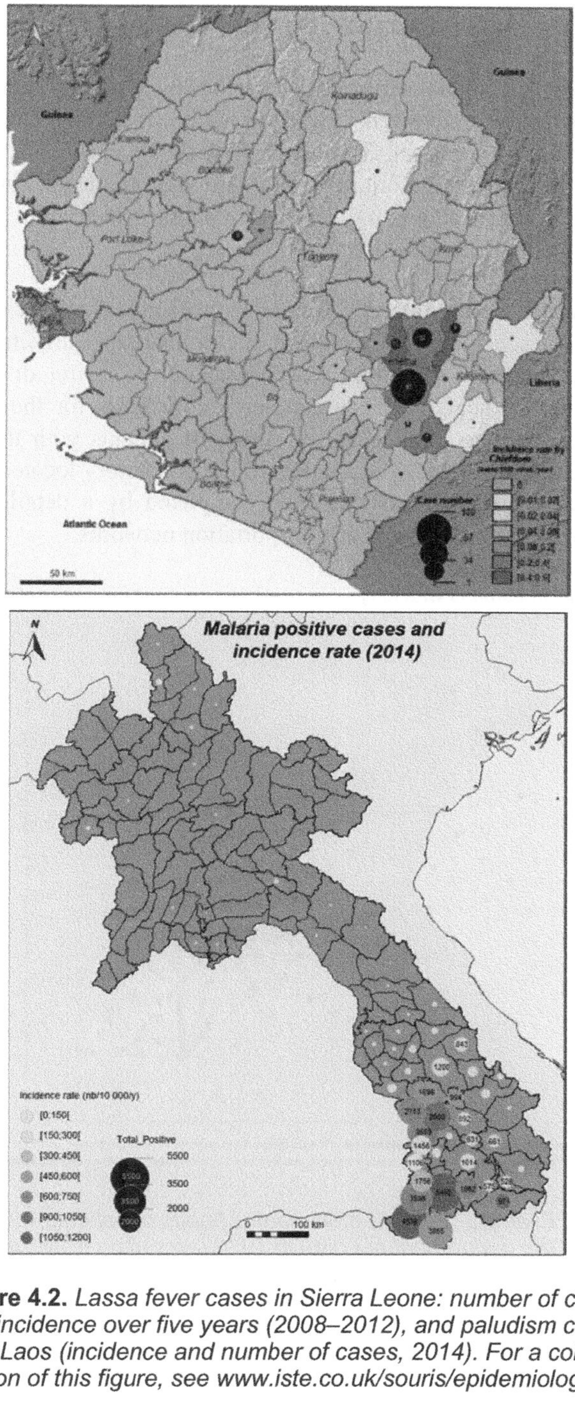

Figure 4.2. *Lassa fever cases in Sierra Leone: number of cases and incidence over five years (2008–2012), and paludism cases in Laos (incidence and number of cases, 2014). For a color version of this figure, see www.iste.co.uk/souris/epidemiology.zip*

4.2.3. *Mapping flows and spatial relationships*

Mapping flows and spatial relationships can be challenging. It generally involves representing patients' movements and visually matching the geographical distribution of patients with the healthcare system. To ensure the clarity of map reading, only the most important flows are represented. Elaborating these representations requires computing spatial selections and aggregations in GIS origin–destination matrix calculation.

In the following, we present an example of this kind of map (called flow map). The map aims to schematically represent the consultation-driven movements of patients with a long-term condition (LTC) residing in Loiret district, in France (Figure 4.3). This medical diagnosis can serve as a basis for the analysis of the matching between the geographical distribution of patients with the LTC and the healthcare system (specialized consultations) on this territory located in the south of Paris urban area. This diagnosis could be completed by a detailed study of the healthcare supply and demand, and of transportation networks.

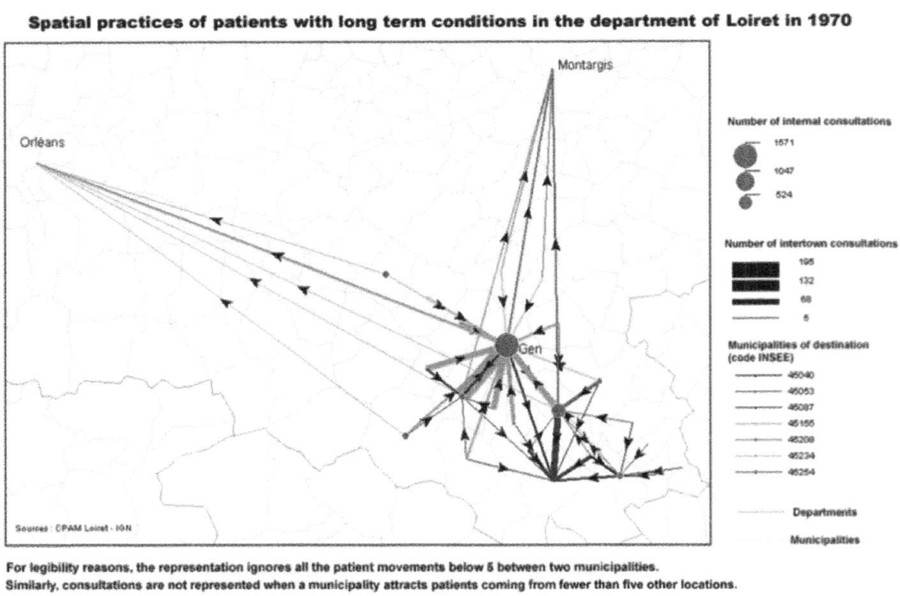

Figure 4.3. *Example of a flow map: spatial habits of patients with long-term conditions in the department of Loiret in 1970 (according to M. Berger, 2007). For a color version of this figure, see www.iste.co.uk/souris/epidemiology.zip*

4.2.4. *Mapping limitations*

Despite having great advantages for epidemiology and health geography studies, mapping has its own limitations, of which we should be aware when interpreting maps, and which can restrict its use. First of all, the use of semiology of graphics or "cartographic language" requires accuracy. Creating a map involves parameters (particularly discretization and colors) that may radically change its interpretation. It is easy to "lie with maps" [MON 93]! Finally, the difficulties inherent to mapping itself must be avoided.

Before further detailing the difficulties in mapping, we will use the classic example of the map representing the prevalence or incidence rate of a disease on a territory. Rates have been calculated for geographical units – geographical areas, for example districts, after having grouped the events into this division, by dividing the number of events in each area by a susceptible population. These areas are colored with a color gradient representing a discretization of rates, for example into equal intervals (Figure 4.4).

The map can be interpreted straight away: to understand what has happened in an area, the reader can detect an overall spatial trend (for example, a north-south gradient), can notice a region where the overall rate is higher or lower (where overall contiguous areas exhibit higher or lower values) and can identify areas with the most extreme rate. How can this visual analysis be described?

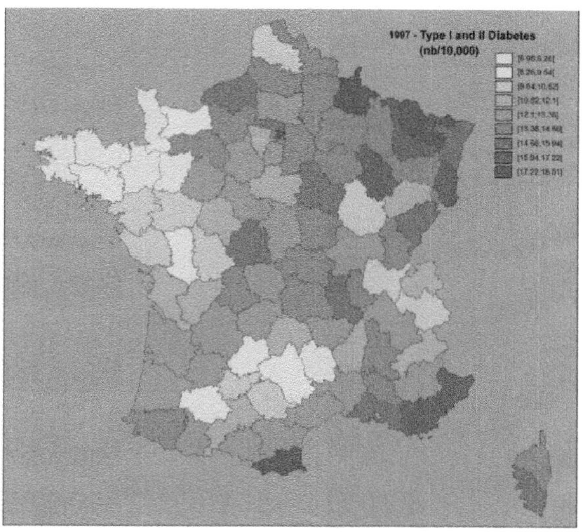

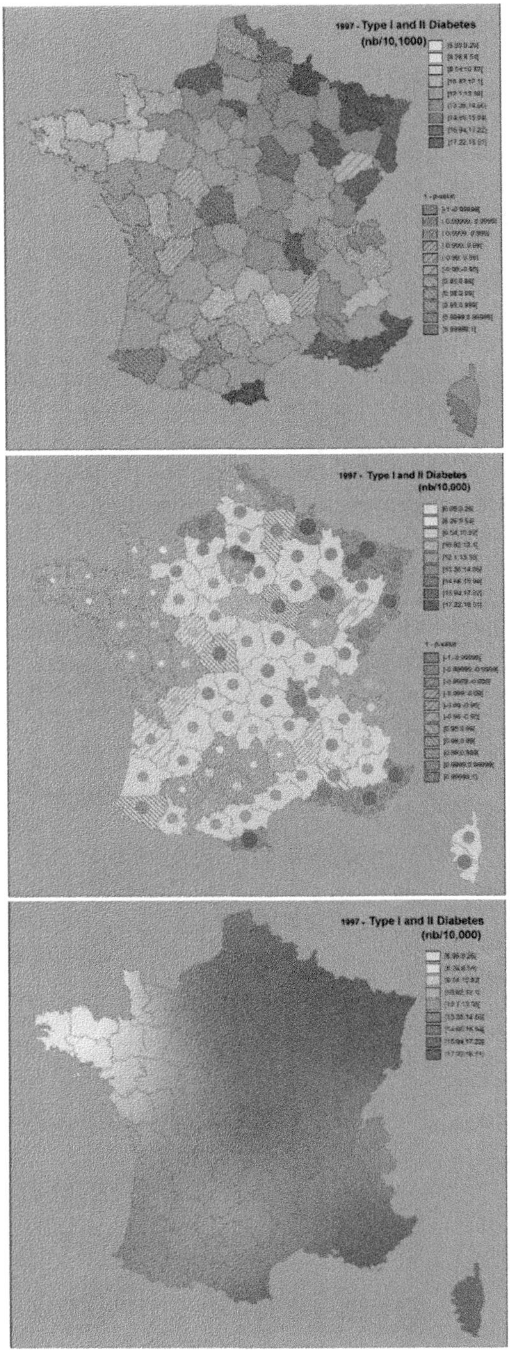
1997 - Type I and II Diabetes
(nb/10,1000)
1997 - Type I and II Diabetes
(nb/10,000)
1997 - Type I and II Diabetes
(nb/10,000)

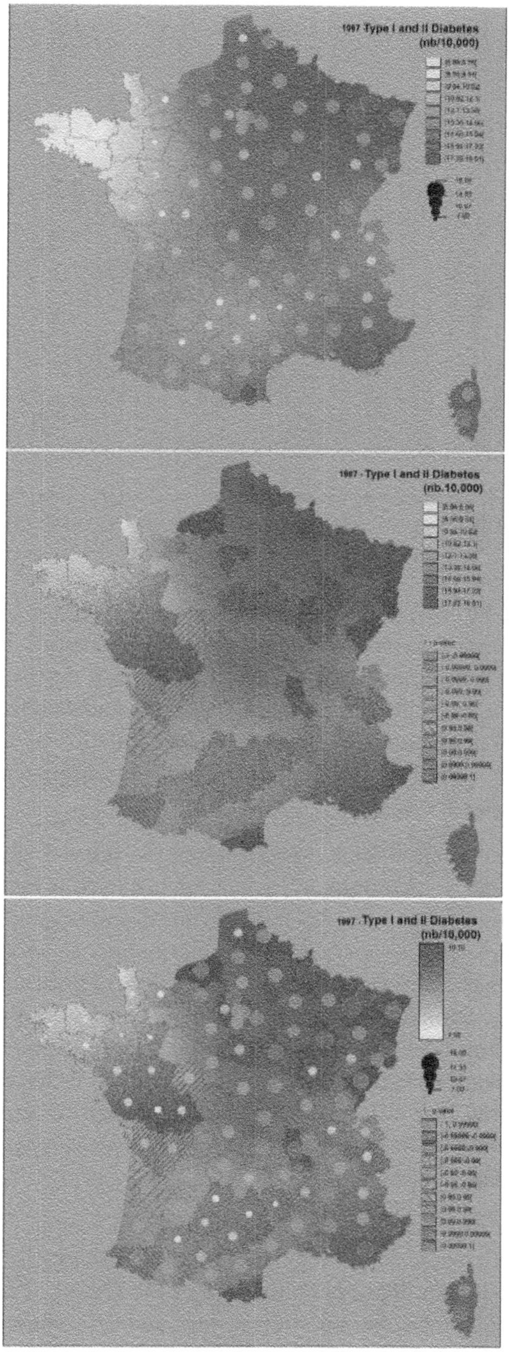
1997 Type I and II Diabetes
(nb/10,000)
1997 - Type I and II Diabetes
(nb.10,000)
1997 - Type I and II Diabetes
(nb/10,000)

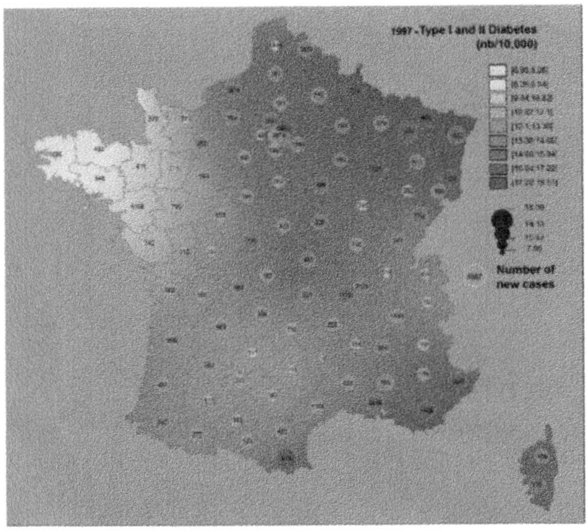

Figure 4.4. *Diabetes in France, 1997, by department. For a color version of this figure, see www.iste.co.uk/souris/epidemiology.zip*

COMMENTS ON FIGURE 4.4.– (1) Incidence representation by a color gradient between minimum and maximum; (2) incidence representation by a color gradient with an indication of the individual significance of each department; (3) incidence representation by a proportional symbol, with an indication of the individual significance of each department; (4) incidence representation by a trend surface after interpolation of the values of the departments; (5–8) various combinations of the previous representations.

– Trying to detect an overall spatial trend is normal and justifiable. The occurrence of a trend is very unlikely; hence, if it exists, it is very unlikely to be random. This situation above all depends on the discretization used, in particular for choropleth maps (see Appendix, available online at: www.iste.co.uk/souris/epidemiology.zip). The discretization of values has significant influence on the visual result of the overall spatial trend. What if, instead of using equal intervals, a discretization based on quantile or deviation from the mean was chosen? Would the situation have been perceived in the same manner? It is necessary to adapt the discretization method used to the phenomenon we want to analyze and highlight (particularly depending on the shape of the distribution of values), without, however, attempting to influence the reader.

– The detection of a large region exhibiting a gradient is also possible, either in excess or in deficiency. However, similar to spatial trends, it depends on the discretization used. It also depends on the size of the region: if the number of areas in the group is small, then the situation is less and less abnormal and can more easily occur by chance.

– Visual detection of a trend, a continuity or a cluster is not the same at the center of the map as on the edge. Topological information (roughly speaking, relationships with neighbors) is richer at the center than on the edges, where the objects have fewer neighbors. This is known as the *edge effect*. Visual interpretation cannot deal with this problem.

– The map represents the various values taken by the "rate" variable over the space under study, and therefore represents relative differences. However, it does not offer any information on the absolute values of these differences. This amounts to noticing the differences while ignoring the overall order of magnitude. Indeed, the map remains identical irrespective of the overall differences – in absolute values – between areas that are either very small or very large. Only the legend can provide this information, and yet we would need to evaluate the distribution of the observed values and then compare this distribution to a distribution resulting from a random allocation of cases to areas. This would be the only way to have an idea of the expected "reasonable" differences and of the homogeneity of the observed values. Hence, it is quite possible to have an unlikely overall distribution (for example, a north–south gradient), although with none of the individual values being considered abnormal. A spatial analysis only focuses on the first point, but quite obviously mapping represents both: the distribution taken as a whole and the individual values of the objects, without usually indicating if they are statistically different or not from what should be expected if the distribution were solely due to chance.

– The map represents the various values taken by the "rate" variable, which is the **ratio of two counts**. A visual analysis above all compares the differences between areas. However, this comparison should be relevant for the problem raised. When the probability of being positive at the individual level is not the same for all the individuals, the count resulting from the aggregation into areas depends on the structure of the population in each area with respect to this probability. If mapping aims to compare rates, then the counts in each area should be adjusted according to this probability – this is called **standardization**. The most common case is age, as many diseases are age-related. Differences in the age structures will consequently be reflected by disease counts, and mapping can primarily highlight these differences in the age structure, which is not what we aim to represent or analyze. Therefore, if we want to analyze differences resulting only from a disease, rather than differences due to age, we need to first eliminate the influence of age on the probability of being sick.

– The map presents the values of a rate, which is the ratio between a number of cases and a population, and implicitly aims to estimate a probability. The accuracy of this estimation depends on the denominator: rate estimation is more accurate if it is calculated on one million individuals than one hundred.

Denominators can differ to a large extent from one area to another, but a choropleth map ignores this information: therefore, the visual comparison focuses on values with different levels of accuracy.

– Here, the map represents areas. The larger the area, the stronger its visual influence. Either the surface is colored depending on prevalence, and the visual influence of a large area will be much stronger than that of a smaller area (which could, however, group the same, or sometimes even larger, counts, as in the urban areas), or the prevalence is represented by a proportional symbol, the visual influence of which is also stronger due to its graphical isolation. Unless there is proper size homogeneity throughout the geographical units (with respect to both area and population counts), any representation involving color should be avoided, and representation by the proportional symbol should be preferred. If a region has units with small areas compared to others (such as urban regions), the symbols are likely to overlap and become illegible: several maps should then be drawn at various scales.

– Other limitations concern the representation of point events. For example, if the spatial support of the phenomenon (where it can occur) consists of well-defined places (such as villages), no event will ever occur outside these places. Yet, cartographic representation does not exclude areas that do not belong to the spatial support of the phenomenon; it represents the entire space. Visual analysis does not address this insoluble problem. For example, a cluster of events on the map can only be observed because the points of the spatial support are clustered (for example, villages along a road or a river) and not because of the phenomenon itself. However, it is very difficult to identify this with a visual analysis. Cartographic representation can easily give a false idea of the clusters of events: these analyses must be backed by specific statistical processing, as will be shown in Chapter 5.

– Finally, clusters of events can also be random and occur by chance: the probability of their presence depends on the number of events and on the position of the spatial support. Visual analysis is tricky, and the presence of a cluster should not be hastily interpreted. As seen previously, the detection of statistically significant clusters is the object of specific statistical processing (Chapter 5).

There are technical solutions to almost all of these limitations. Some of them can be addressed at the level of cartography as indicated above (nevertheless, one can find maps that do not take them into account), while others must be mentioned in the notes and must be backed by additional statistical and spatial analyses, which will be part of the general analysis of the studied phenomenon.

4.2.5. *Mapping rate significance*

On a choropleth map, the representation of rate alone does not indicate the absolute magnitude of the values; the reader must refer to the legend to get this information, and the notice should indicate the magnitude of the values we would observe if the spatial distribution was random. It may therefore be useful to superimpose different types of information on the same map: number of cases (by a proportional symbol, displaying the absolute value), rate of incidence or prevalence (classified and represented according to a color scale) and individual significance of the observed/expected ratio.

We are within the framework of data aggregated by geographical units. One way to graphically account for the extent of a number of cases or of a rate in absolute value is to map not only the observed value, but also its significance with respect to an *a priori* distribution (depending on what is known about the phenomenon; in the absence of hypotheses, it is a random distribution). Significance refers to the notion of expected count and of the confidence interval of this count or of the observed/expected ratio. This significance depends on the ratio and size of the population, and must be calculated for one geographical unit at a time (Breslow and Day's test [BRE 87, BOU 93]). It must be expressed in terms of the null hypothesis and the alternative hypothesis.

Suppose that the probability p of a disease in a population is accurately known, and identical for each individual (for example, equal to 1/10,000). This probability is therefore identical in all geographical units of the territory, and the differences between units are not random. According to the binomial law, the standard deviation of the expected number of cases in one unit depends on the population n of the unit and on the individual probability: $v = \sqrt{\frac{p(1-p)}{n}}$. The confidence interval can thus be calculated for each unit.

The overall population is 1,000,000: two large units consist of 100,000 people each, while the other 98 units consist of the rest of the population, that is, over 8,000 persons per unit. Let us now consider a randomly simulated situation in this population. By definition, the differences in the number of cases between geographical units can only be due to random variations. The following results are obtained: one of the large units has 10 cases; the other one has 11; most of the others have no cases at all, some have one; three of them have two and one has three. The differences between units are not statistically significant: these results are compatible with the random distribution.

Conversely, relying on values observed per unit in order to calculate various probabilities per unit means challenging the *a priori* hypothesis of an identical probability for each individual, independently of its unit. This questioning is not valid unless the observed distribution is incompatible with a random distribution (modeled by the binomial law or Poisson law which allows its approximation).

Box 4.4. *Differences are not always statistically significant*

Significance is individual, but the number of units that are significantly different from their expected value depends on the overall number of units: this is a multiple test problem. Mapping significance must reflect the problem raised in the note in the map:

– In the case of individual significance, all the units that individually reject their null hypothesis can be highlighted.

– In case of overall significance, the overall null and alternative hypotheses must be specified from individual hypotheses. The expression of overall hypotheses must consider that the more geographical units there are, the higher the probability of obtaining extreme values solely by chance. In case the overall hypothesis retains the same risk of error (for example, 5 %), the threshold for rejecting the null hypothesis should be changed for each unit (for example, with the Bonferroni correction, which involves dividing the risk of error by the number of tests conducted). It is also possible to change the overall alternative hypothesis (for example, choosing to reject the overall null hypothesis if at least two individual units reject their null hypothesis). Mapping the units that are involved in the rejection of the overall null hypothesis requires clarification of these choices in the note to the map.

4.2.6. *Rate adjustment*

The discussion remains within the framework of data aggregated per geographical unit, with a rate perceived as the estimator of probability per geographical unit.

The problem discussed in the previous section can be addressed solely from the point of view of the accuracy of the estimation of the rate per unit. Indeed, when the implicit objective of the case/population ratio is rate estimation, the count of the denominator determines the estimation accuracy: the larger the counts, the higher the estimation accuracy. When this estimation is done for a set of areas, estimation accuracy is not the same for all the areas, since it depends on the area count.

Suppose a rate constant and equal to 0.3 over a set of geographical units. There are two susceptible subjects in one unit, and 100 in the other. The unit consisting of two people can only take the values 0, 0.5 or 1. Therefore, estimating 0.3 by one of these values can only be far less accurate than in the second unit, where the accuracy of the possible values is 0.01. Subsequently, the rate estimation using observed values does not have the same validity in the two geographical units, both in terms of calculation accuracy (related to the count) and of random variability (which depends on count and also on the actual rate). For accuracy reasons only, the value observed in the first unit is much farther from the actual value than that observed in the second unit.

The lower the rate, the more difficult it is to estimate the probability of observed values. According to the aforementioned example, the rate calculation using observed data leads to significant differences between the geographical units. The two large units have rates of 0.00010 and 0.00011 (high accuracy calculation), the majority of units have a rate equal to 0, those with one case have a rate of 0.000125, those with two cases have 0.00025 and the one with three cases has 0.000375. These differences are solely due to random variations and differences in the denominator, since rates were assumed to be independent of location. Thus, we should avoid the use of these rates to estimate the actual rate in one unit (by hypothesis considered equal to 0.0001 in all the units).

Box 4.5. *Estimating probability with observed values*

If the objective is to estimate a risk (a probability), mapping the various rates observed in the units may mislead the reader; mapping inevitably leads the reader to compare values, and the information on the denominator, and thus the accuracy, are not represented. Quite often, the geographical units exhibiting extreme values are precisely those with the smallest populations, and therefore with the least accurate rate calculation. In these units, the lack of accuracy is far more significant than the random variability of the units with larger population, where accuracy is more significant. Certainly, the problem goes far beyond cartography and concerns any rates comparison used for public health purposes and the intervention of health authorities. We will discuss this issue further in Chapter 6, where risk assessment methods will be examined.

Various rate adjustment techniques can be used in order to reduce the visual effect of accuracy differences due to differences in counts. The general idea is to collect information outside a given unit in order to improve the accuracy in that unit: the underlying hypothesis is that the studied area has the same characteristics as a larger area.

The first adjustment technique is the empirical Bayes estimators (EBE). This uses the characteristics of the whole studied population. An overall average rate and its variance are estimated for the population as a whole in all the geographical units. To do this, an *a priori* hypothesis on the shape and parameters of the distribution must be formulated, followed by an empirical estimation of this shape and of these parameters depending on the observed values [CLA 87, MOL 99]. Adjustment is then made by using a linear combination of the observed value and the *a priori* value; the coefficients depend on the relative size of the population and the *a priori* variance in the unit. Hence, the lower the count of the unit and the farther the rate value from the overall rate, the more significant the adjustment.

This adjustment technique is not exclusively used in cartographic representation: it is also used in the spatial analysis of rates, particularly for autocorrelation indices (see Chapter 5).

Let λ_i be the unknown rate in a geographical unit i and $r_i = y_i/n_i$ the observed rate, where y_i is the count observed and n_i is the size of the population of unit i. Let us make the hypothesis of an *a priori* distribution of probability for each unit, of mean μ_i and variance φ_i. The best Bayes estimation $\hat{\lambda}_i$ of λ_i is the result of a linear combination of the *a priori* distribution and the observed data, with coefficients w_i and $(1-w_i)$ which, depend on the size of the population and the variance of the *a priori* distribution:

$$\hat{\lambda}_i = w_i r_i + (1 - w_i)\mu_i \text{ with } w_i = \frac{\varphi_i}{(\varphi_i + {}^{\mu_i}/_{n_i})}$$

In order to calculate w_i, a hypothesis on *a priori* distributions must be formulated. For example, let us suppose that the means and variances are identical for all the geographical units, and that they follow a Gamma distribution (mean $\mu = \nu/\alpha$, variance $\psi = \nu/\alpha^2$). Then, μ and ψ can be empirically estimated from the data observed on the set of units:

$$\hat{\mu} = \frac{\sum y_i}{\sum n_i} \text{ and } \hat{\psi} = \frac{\sum n_i (r_i - \hat{\mu})^2}{\sum n_i} - \frac{\hat{\mu}}{\bar{n}}$$

The Bayes estimator $\hat{\lambda}_i$ of the rate λ_i is then:

$$\hat{\lambda}_i = \hat{\mu} + \frac{\hat{\psi}(r_i - \hat{\mu})}{(\hat{\psi} + \frac{\hat{\mu}}{n_i})}$$

Box 4.6. *Empirical Bayes estimator*

SavGIS	The empirical Bayesian adjustment is available with the following menu: Calculate → Statistical calculation → Bayesian adjustment
R	The adjustment calculation is available with EBest and EBlocal procedures in the spdep library
GeoDa	The calculation of Bayesian adjustment for the rates is available with the menu: Map → Rates-Calculated Map → Empirical Bayes

Table 4.2. *In practice*

Another adjustment technique involves making a spatial smoothing to minimize the differences between neighboring units. A hypothesis on the spatial autocorrelation of rates is formulated (see Chapter 5), which allows the adjustment of less accurate values that are subjected to higher random variability with more accurate neighboring values.

We can also mix the two techniques, which, instead of calculating the parameters of the *a priori* distribution on the set of areas, only calculate the parameters of the neighboring areas (in terms of distance or contiguity). Geographical units can also be classified into several groups (based on geographical criteria, such as urban–rural). The *a priori* distribution is then chosen and its parameters are calculated class by class, considering only the units of the class.

It is also possible to use kernel interpolation techniques (in two dimensions), with a variable radius in order to integrate enough individuals of the neighboring units for a proper rate evaluation by interpolation [LAR 11].

Finally, the problem can also be solved with an approach similar to the one of significance. When estimation accuracy is low and subjected to strong random variation (which is often the case for rare events), rate calculation in a unit is no longer concerned with the accurate estimation of the actual value of risk in this unit. Instead, it seeks to determine whether the unit value actually differs from the overall value, yielding a quantitative result solely in the form of a classification reproducing the inaccuracy of the estimation. Then, cartography directly uses this classification, rather than discretizing the rate in order to represent it [ABR 13]. The same is true for decision-makers in public health, who very often rely on the risk classification, instead of the actual value, either adjusted or not. Rather than making a classification *a posteriori* by a discretization of the observed rates, a classification is here the objective of the estimation and adjustment procedure.

4.3. Descriptive statistics and visual synthesis tools

Visual synthesis tools are used to replace a set of objects with only one object synthesizing a spatial characteristic of the set of objects, such as centrality, extent, dispersion and isotropy (invariance with respect to direction).

4.3.1. *Average points, median points*

The **mean point** is the center of gravity of the cloud of points. It results from the operation in two dimensions that corresponds to the mean in one dimension. Similarly, the **median point** corresponds to the median. The mean point or the median point can be used to synthesize the localization of a set of points by only one point, which represents the "center" of this set:

– The mean point is calculated as the average of the coordinates of the points on each axis. It has the property of minimizing the sum of squared distances to all the points of the set. It can be easily calculated, but has several drawbacks: it depends on the system of reference (axes), and its position is more sensitive to the distant points than to the nearby points. There are several ways in which the calculation can be improved in order to avoid these drawbacks. We can take the mean point of the set of mean points calculated by rotating the axes around the initial mean point; then, we can calculate a mean point by weighting the calculation by the inverse of the distance of the points to the initial mean point.

– The median point minimizes the sum of distances to the points in the set. Considering the so-called Manhattan distance ($d(P,Q) = |xp - xq| + |yp - yq|$), the coordinates of the median center can be readily deduced from the coordinates of the points. Considering the Euclidean distance, the problem has no analytical solution and can only be solved by successive approximations. It does not depend on the system of reference and it is less sensitive to distant points (Figure 4.5).

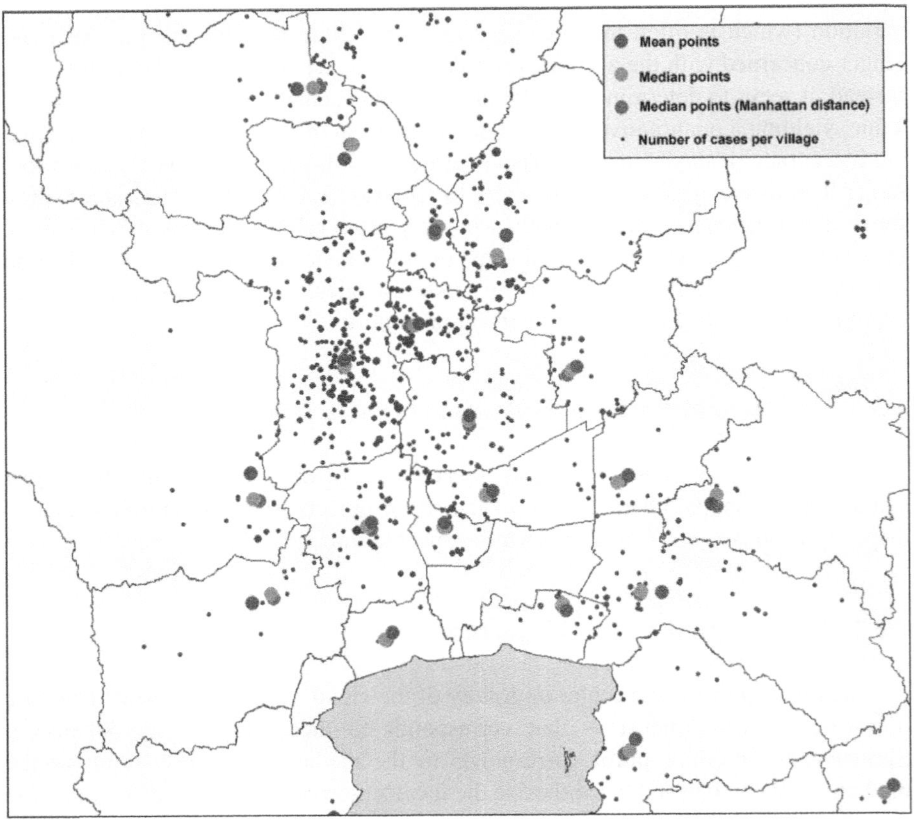

Figure 4.5. *Mean points and median points, per province, calculated from cases localized in villages (avian influenza, Thailand, 2004–2008). For a color version of this figure, see www.iste.co.uk/souris/epidemiology.zip*

Similar to a center of mass, the cloud points can be weighted, assigning a weight to each point, and the center of the spatial distribution is then calculated by considering the relative significance of the points given by their weight.

The mean or median point of a group of objects can be used as localization for mapping with a symbol a value of this group of points. It can also be used in building indices aimed to express a geometric characteristic.

Other geometric constructions can be used to account for centrality in a point cloud: the center of the circle inscribed in a convex envelope and the center of the circumscribed circle.

SavGIS	The calculation of mean and median points is available with the menu: **Type → Creation of points → Mean points**
ArcGIS	The calculation of mean and median points is available in ArcToolBox: **Spatial Statistics Tools → Measuring Geographical Distributions → Median Center**

Table 4.3. *In practice*

4.3.2. *Standard deviational ellipses*

By analogy with standard deviation, standard distance is defined as the square root of the mean of squared distances of the points $P_1, \ldots, P_n$ of the point cloud to the mean point M:

$$SD = \sqrt{\left(\frac{1}{n}\sum_{i=1}^{n} d^2\,(M, P_i)\right)}$$

Standard distance offers an indication on the extent of a set of points [BAC 63].

Another objective is to account for the mean orientation of the cloud of points. To do so, we first define a first axis that maximizes the sum of distances between the projections of the points on this line and the (mean or median) center also projected on this line. The dispersion angle is defined as the angle between the x-axis and this first axis; the main standard distance is defined as mean of the distances between the points and the center projected on the first axis; and the secondary standard distance is defined as mean of the distances between the points and the center projected on the perpendicular to the first axis.

These three different elements can be synthetically visualized by an ellipse, called the standard deviational ellipse (SDE): its center is the mean center; its major axis is oriented as the first axis and corresponds to the standard distance of the points projected on the first axis; its minor axis corresponds to the standard distance of the points projected on the line perpendicular to the first axis.

Standard deviational ellipses can be used to visually compare the spatial distribution of several sub-groups (Figure 4.6).

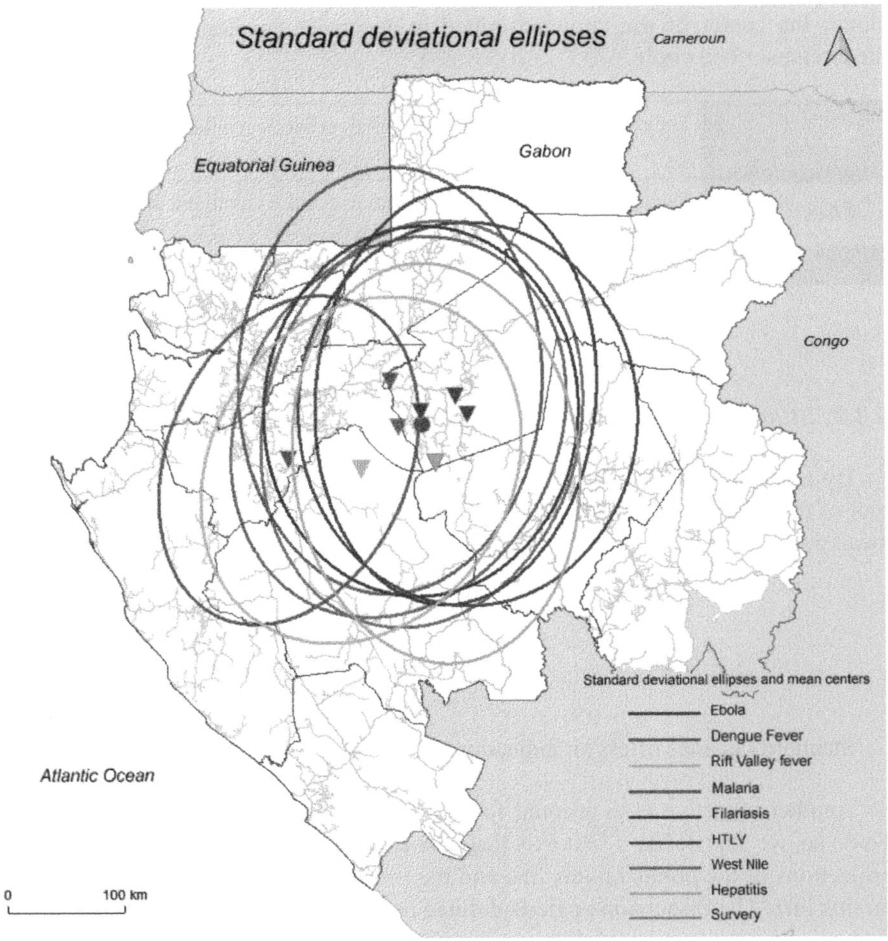

Figure 4.6. *Standard deviational ellipses used in order to compare the spatial distribution in a seroprevalence survey of several pathogens (Gabon, 2010). For a color version of this figure, see www.iste.co.uk/souris/epidemiology.zip*

SavGIS	Standard deviational ellipses can be created with the menu: Type → Creation of areas → Standard deviational ellipse
ArcGIS	Standard deviational ellipses can be created in ArcToolBox: Spatial Statistics Tools → Measuring Geographical Distributions → Directional Distribution (Standard Deviational Ellipse)
R	Standard deviational ellipses can be created with the calc_sde function of the Aspace package

Table 4.4. *In practice*

4.4. Interpolations and trend surfaces

When the cartographic representation focuses on evidencing spatial trends or patterns, specific techniques can be used in order to advance this objective and eliminate some interpretation problems. Indeed, initial objects are not necessarily evenly distributed in space and their spatial distribution can "pollute" or hide the overall trend. The reader of a thematic map is naturally inclined to search overall spatial trends, in the sense of spatial distribution without local variations or noise: a gradient, an alignment, a privileged direction, a cluster, a centrality, etc. The spatial trend is a shape or a simple geometric characteristic that allows the modeling of a complex surface in a three-dimensional space.

Rather than directly representing the objects (with symbols or colors for each object), the purpose of mapping is to represent the variable (in the numeric case) or the counts (in the Boolean case) with a continuous surface obtained by interpolation or estimation over the studied space. This surface consists of values on all the studied space, even when there are no initial objects. The purpose of the calculation is the cartographic representation, and not the result itself. Its interpretation must take into account these limitations.

Two different approaches are used in order to calculate this surface: on the one hand, an interpolation-based approach, and on the other hand, a modeling-based approach.

4.4.1. *Interpolations and continuous representation*

The first solution involves the interpolation-based calculation of values on a grid of points uniformly distributed on the studied space. The graphic representation of this continuous surface facilitates and supports the visual detection of gradients or clusters. Various image representation techniques can be used to represent this surface: color gradient, wireframe representation and perspectives (Figure 4.7).

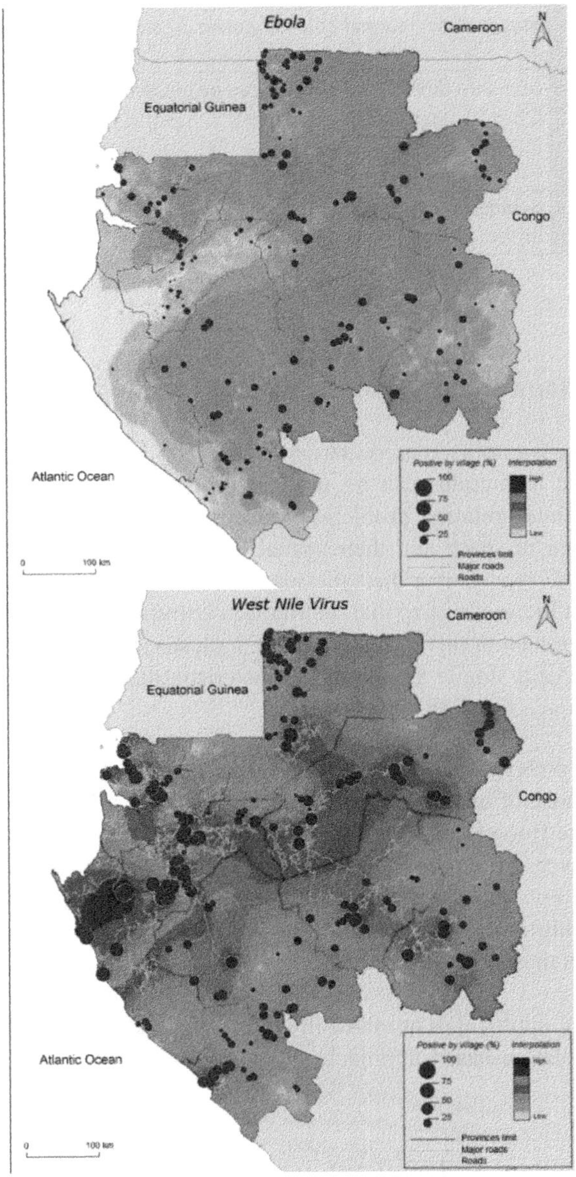

Figure 4.7. *Prevalence rate for two antibodies (Ebola, West Nile virus) in the rural population in Gabon, studied on a sample of villages. The values represented by proportional symbols on the sample allow only with great difficulty the detection of overall tendencies, in contrast with the trend surfaces calculated by interpolation, which clearly show the differences between the two distributions, but which represent values in the forested areas where there are no inhabitants and no surveyed villages. For a color version of this figure, see www.iste.co.uk/souris/epidemiology.zip*

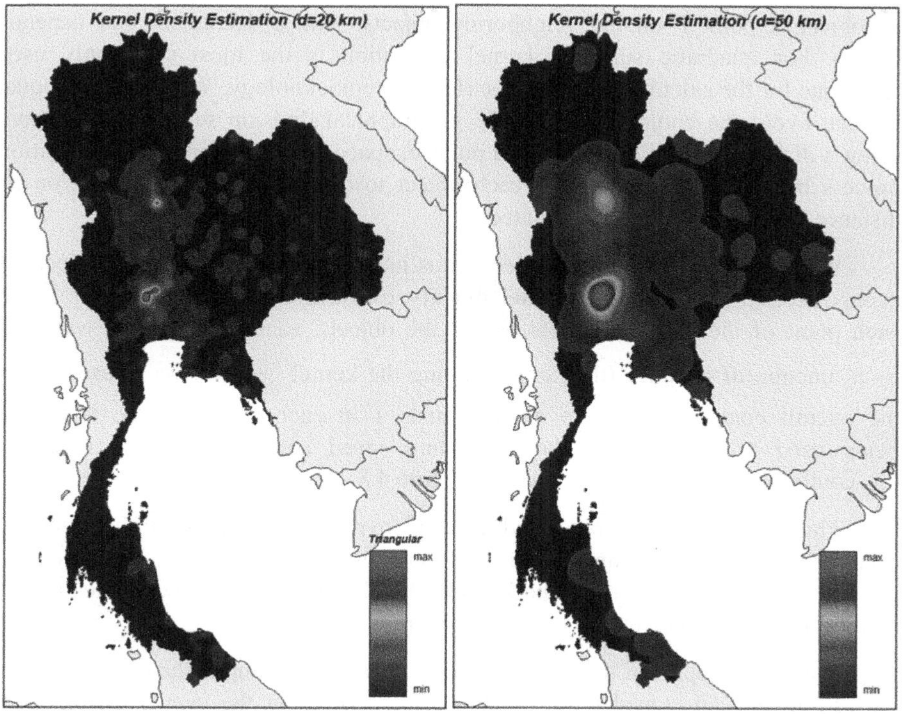

Figure 4.8. *Density of events (farms infected with avian influenza in Thailand, 2004–2008). Kernel density estimation (Gaussian), range: 20 km (left) and 50 km (right). For a color version of this figure, see www.iste.co.uk/souris/epidemiology.zip*

Interpolation, estimation and smoothing techniques are used in order to calculate the value at each point of the grid, depending on the values or the presence of objects. The raster unit depends on the fineness that the graphic representation is intended to have. Several techniques are used:

– When the variable to be mapped is numerical, **kernel estimation** calculates for each point of the grid the weighted mean of the neighboring values weighted by a function of distance (kernel), for all the objects located at a distance below a given distance *d*0 [BOW 97]. The kernel, a decreasing function of the distance *d* of the object from the point of the grid, can be a linear function (for example, $(d0 - d)/d0$), a quadratic function (for example, $\left(\frac{d0-d}{d0}\right)^2$) or a Gaussian function (for example, $e^{-1/2(\frac{d}{d0})^2}$). It is like spreading the value of an object at point P on the neighboring objects, with a decrement that depends on *d*0. For quadratic surfaces, kernel estimation is the most commonly used technique for the calculation of trend surfaces in epidemiology. With this technique, we can avoid the choice of a specific geographical division for the aggregation, using a disk of radius *d*0 around each point in space, and enriching the aggregation process by assigning a weight to each object to be aggregated, depending on its distance from the point to be estimated.

– When the variable to be mapped is qualitative or when a density of objects or events needs to be mapped, **kernel density estimation** requires summing up at each point of the grid the counts for all the objects, each object being weighted by a function of distance (the kernel). Using the kernel $\frac{1}{\sqrt{2\pi}}e^{-1/2(\frac{d}{d0})^2}$ ensures that the overall contribution of an object equals 1 in each direction, the factor *d*0 being used to parameterize the decreasing speed and therefore the range of this contribution depending on distance (Figure 4.8).

– This technique can also be used to calculate rates at each point of the space, by dividing the weighted counts of the phenomenon by the weighted counts of the overall population, using the same kernel. The above-mentioned techniques (EBE) (Figure 4.9) can also be used to adjust these rates.

– When the support of the phenomenon is linear (for example, the road network), the estimation can consider the length of the support in the events' search disk (NKDE) [XIE 08].

Other interpolation techniques are also available for estimations for any point in space:

– Potential-based interpolation, which calculates for each point of the grid the mean of the values of the objects weighted by a function of the inverse of the distance of the objects from the point of the raster.

– Barycentric interpolation, which calculates at each point of the raster the mean of the values of neighboring objects – in the sense of Voronoi – weighted by the inverse of the distance between the object and the point on the grid.

– Bézier surface interpolation (and more generally piecewise polynomial interpolation, such as spline, the polynomial coefficients being calculated from the values of the objects).

– Geostatistical interpolation (for example, kriging, a linear interpolation method that ensures minimum variance).

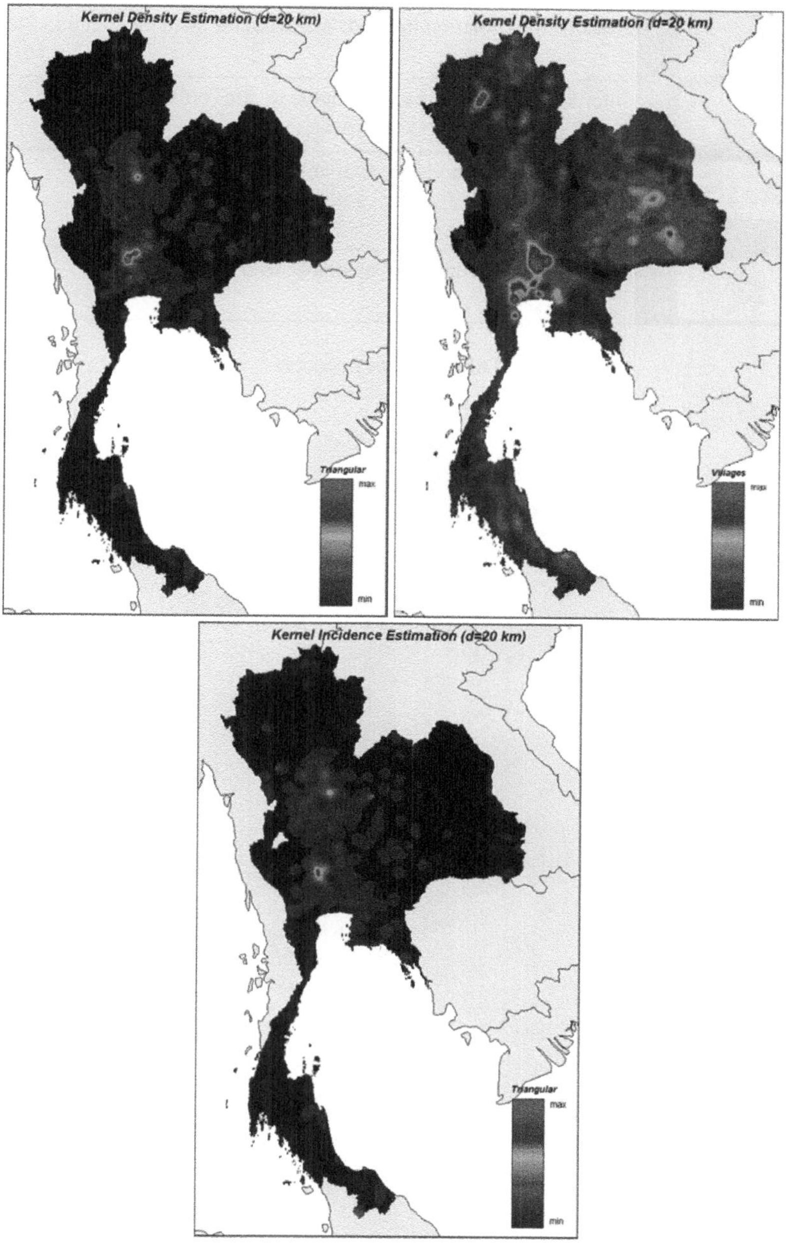

Figure 4.9. *Kernel density estimation (Gaussian) at each point. Maximum research radius: 20 km. On the left: density of farms infected with avian influenza. At the center: density of villages. On the right: incidence of infected farms. For a color version of this figure, see www.iste.co.uk/souris/epidemiology.zip*

SavGIS	Various interpolation methods are available with the menu: Babel → Interpolation
ArcGIS	Interpolation methods are available in ArcToolBox: Spatial Analyst Tool → Interpolation
QGIS	Interpolation methods are available with the menu: Processing Toolbox → SAGA → Geostatistics → Raster Creation Tools
R	Interpolation methods are available with gstat package.

Table 4.5. *In practice*

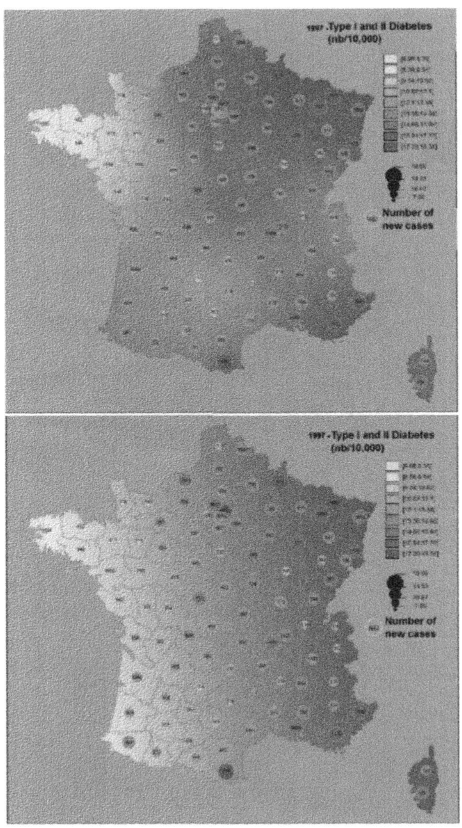

Figure 4.10. *Trend surfaces for the incidence of diabetes in 1997 in France. On the left, a representation by interpolation (potential); on the right, the trend plane of first-order (z = 7.55 + 0.003461*x + 0.000778*y). The orientation of the bilinear trend is here a gradient slightly oriented south-west and north-east (13°)*

4.4.2. *Directions and gradients*

The second solution for trend surface evaluation involves the direct search for a surface by statistical modeling, using only functions of geographical coordinates as explanatory variables (either the coordinates as such, or their products or powers). This surface is generally much farther from reality than the surfaces resulting from interpolation.

> A trend surface is a statistical modeling of a numerical variable of objects depending on their coordinates (Figure 4.10). In general, the chosen model is a polynomial function of coordinates. The first-degree coefficients represent the plane that minimizes the sum of distances from the observed values.

Box 4.7. *Trend surface*

SavGIS	The calculation of a trend surface is available with the menu: Calculate → Statistical calculation → Multivariate modeling After having calculated the coordinates of each object with the menu: Calculate → Coordinates

Table 4.6. *In practice*

4.4.3. *Anamorphoses*

An anamorphosis is a cartographic representation built by deliberate distortion of space from a quantitative variable. It is most often used to expressively highlight a phenomenon across a country or over a planisphere [BRU 87] (Figure 4.11). It is used by the media in order to draw the reader's attention.

SavGIS	The calculation of an anamorphosis is available with an external module (ScapeToad, for example), after having exported the objects in Shapefile with the menu: Frame → Export → In Shapefile The Shapefile generated by anamorphosis can afterwards be imported with the menu: Frame → Import objects → localized → From a Shapefile

Table 4.7. *In practice*

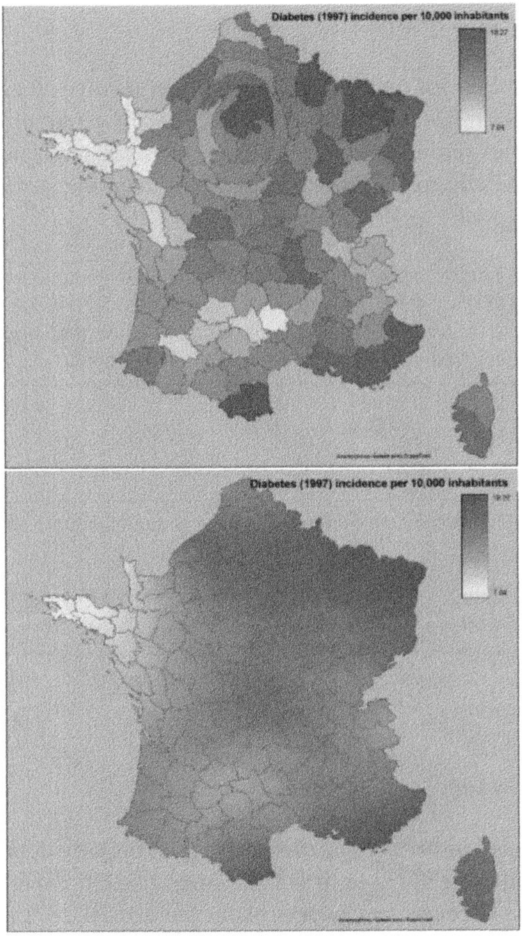

Figure 4.11. *Incidence of diabetes in 1997 in France. The Paris region is greatly magnified, as its departments have large values but are small in size. Due to anamorphosis, their size becomes comparable to that of more extended departments of similar value, while the size of low-incidence departments is reduced*

4.5. Spatio-temporal animations

4.5.1. *What and how*

Time-localized data can be used to show the temporal evolution of a parameter. Cartography aims to represent either the occurrence of an event, the number of events per geographical unit within a temporal window, or an index calculated with the events per geographical unit within a temporal frame (prevalence, incidence, risk, etc.). It is thus common to produce many maps presenting the retroactive evolution of a phenomenon.

The graphic elements represented are either related to events – positions, values and indices, synthesis positions (mean centers or mobile centers), geometric characteristics (clusters, centrality) – or to movements – movements themselves (arrows), or events related to the same individuals (arrows), or sequences of events (arrows representing a pathway; see Figure 4.12).

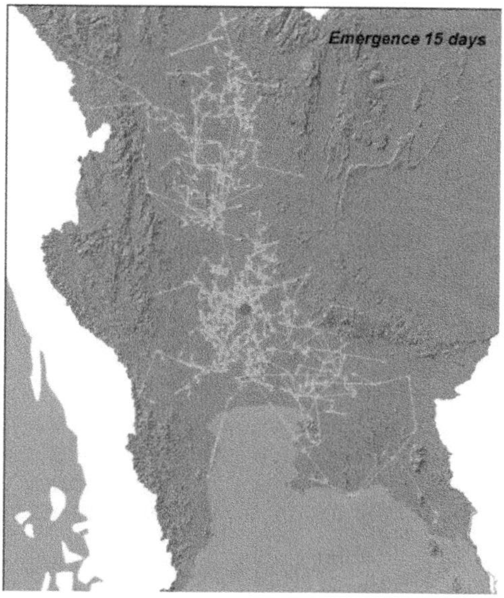

Figure 4.12. *Mapping of the supposed pathway of the avian influenza outbreak in Thailand (2004), from an emergence point (index case) [SOU 10]*

SavGIS	Arrow mapping is available in the cartographic browser and serves to represent the layer of lines created for modeling a pathway.

Table 4.8. *In practice*

4.5.2. *Animated mapping*

The next stage involves the generation of animations from static maps. It is thus possible to visualize the occurrence of cases per unit of time; the number or the density of cases per unit time and per geographical unit (that can also be distorted in order to obtain animated anamorphoses); the movement of the mean center; the movement of aggregates; and the follow-up of sites of concentration (animated heat maps). Animation can be readily generated from a series of maps using standard and free software programs (for example, Windows Movie Maker developed by Microsoft). On the contrary, it is difficult to present animations in an article or book (Figure 4.13).

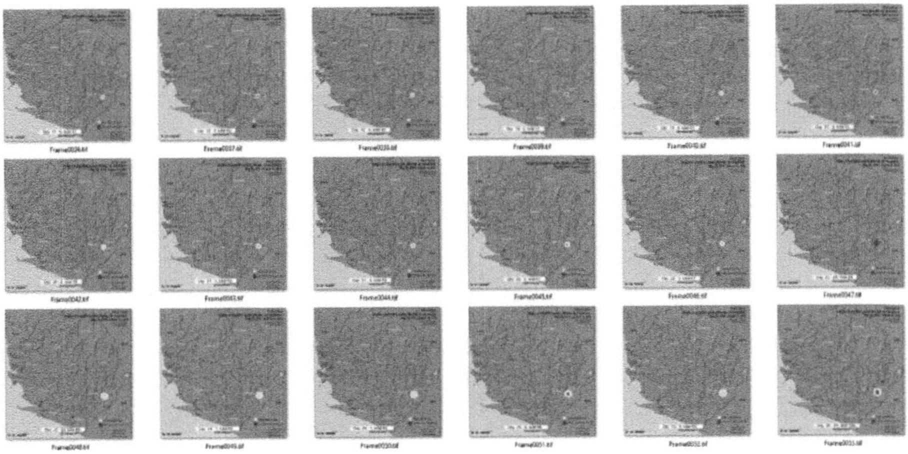

Figure 4.13. *Images extracted from an animation showing the day-by-day evolution of the Ebola outbreak in Sierra Leone (number of cases by county seat, 2014). New cases are represented in red; aggregated cases are represented in yellow. For a color version of this figure, see www.iste.co.uk/souris/epidemiology.zip*

SavGIS	Temporal animations can be generated with the menu: Frame → Temporal animation
GeoDa	Temporal animations can be generated with the menu: Tool → Map Movie
QGIS	Temporal animations can be generated with the plugin: Time Manager (adds controls for time management in QGIS)

Table 4.9. *In practice*

Animation is a source of further difficulty, besides the mapping difficulties that have been presented in this chapter. The rules of mapping and of the semiology of graphics must be observed.

SUMMARY.–

– Mapping allows the visual analysis and detection of overall characteristics (trends, centrality, patterns, clusters, periodicity, etc.).

– Mapping involves many rules that are stated using the terminology of the semiology of graphics.

– Mapping of rates is commonly used, but precaution is recommended. Several statistical techniques for adjustment allow the minimization of the influence of differences in accuracy in the calculation of rates.

– Standardization is required in order to eliminate the influence of known risk factors that must be omitted in the representation.

– Visual synthesis tools are available for the representation of the geometric characteristics of the spatial distributions of clouds of points.

– Interpolation allows the representation of discrete phenomena by continuous surfaces that are used to represent spatial tendencies.

5

Spatial Distribution Analysis

5.1. Introduction

Spatial analysis refers to any analysis using the localization of the studied individuals. There are many methods for conducting spatial analyses. This chapter, along with the following ones, presents the main methods used in epidemiology and health geography.

This chapter presents the spatial analysis methods known as *direct methods*, which are used for analyzing the spatial distribution of objects or of their values directly from their localization. Chapter 6 will present spatial analysis methods that use the spatial aggregation of values into geographical units.

5.1.1. *"Direct" methods for spatial analysis*

The formalism described in Chapter 2 is used here: domain of definition D, function F. The direct methods presented here are classified as follows:

– global analyses, which provide information on the global spatial distribution (that is on function F, considered over its entire domain of definition D, and its global trends);

– local analyses, which provide information on the local variations of the spatial distribution;

– detection of places in space (points or surfaces, like hot spots or clusters) that present a specific character with respect to the global spatial distribution of F;

– estimation and interpolation, which allow the characterization of any point in space with respect to the spatial distribution of known values of F.

It is worth recalling that the purpose of all these analyses, either global or local, is mainly to decode and understand the global process (or processes) that have generated the spatial distribution of a health phenomenon.

Many methods of analysis use the principle of *statistical testing*. The problem to be solved is classically expressed by stating two hypotheses: a null hypothesis, H_0, and an alternative hypothesis, Ha. The principle is to compare, by means of a numeric index that synthesizes the problem to be solved (called a *test statistic*, denoted *S*), the observed situation to the set of possible situations under H_0. The observed situation is considered to be a realization within the set of possible situations given H_0. When the index of the observed situation is among those of situations whose probability is below a chosen threshold, H_0 can be rejected and Ha can be accepted, with a risk of error corresponding to this threshold (the significance level of the test).

The evaluation of the set of all possible situations under H_0 and the statistical distribution of index S under H_0 is essential. It is a function of H_0 given *a priori* on the phenomenon (random, "contagious", uniform spatial distribution). In most cases, H_0 corresponds to a random situation (which amounts to formulating no *a priori* hypothesis on the phenomenon).

A test statistic must synthesize the problem to be solved or the characteristic to be highlighted. The index is a numeric value that can be ordered according to the problem to be solved and the characteristic being studied.

When the test concerns a spatial characteristic, the problem is geometric in dimension 2 or 3, and the index S cannot be a direct measure or the occurrence number of an event, as is the case in dimension 1. The index is therefore built as a numeric function, based on geometric or topological characteristics, in order to synthesize a geometric problem in one dimension, formulated in two or three dimensions. Two types of methods are used for the construction of these indices:

– methods that directly use the location of objects, the spatial relationships between objects (distances, contiguity, adjacency) or the differences in value between objects. For example, the analysis of the clustering of objects may use the minimum distance between an object and its neighbors (nearest neighbor methods) or a calculation involving the relationship between distance and value (spatial autocorrelation indices, see below);

– methods that use the counting of objects, events or the mean of an attribute within geographical windows (disk, square unit cell, hexagonal unit cell, etc.) that cover the studied space, or within a window moving over the studied space (for example, spatial scan methods or functions derived from Ripley's K-functions).

The statistical distribution of the values of these indices under H_0 is very often obtained by simulation (known as the *Monte Carlo* simulation in reference to the equiprobability of the simulated events), as the mathematical determination of the theoretical distribution of the index may be very difficult, particularly when indices derive from geometrical calculations. Simulation is used in order to generate a large sample of possible situations under H_0, and thus provides the possibility to approach the distribution of values of the index without having to consider the spatial distribution of the set of objects, particularly when they are points. It also allows the solving of edge effects.

The use of simulation to estimate the distribution of possible values of an index replaces an often difficult mathematical reasoning, notably taking into account the location of all objects and the edge effects. This experimental approach involves a few constraints, but must have solid bases, particularly with respect to the definition of possible situations and methods employed for generating these situations. The approach is always the same: formulation of hypotheses, definition of a numeric index that discriminates with respect to the problem to be solved, simulation of possible situations, calculation of the index for each simulated situation, estimation of the parameters of the index distribution, comparison with the observed situation and, if the null hypothesis is rejected, estimation of the probability observing values of the index at least as extreme as the index of the observed situation (p value).

The term edge effect denotes the fact that topological or metric information is not the same at the center as at the edges, where objects have fewer neighbors. Therefore, the quality of an estimation that uses the value of neighbors is not the same on the edges, because there is less information. Monte Carlo simulations make it possible to consider the shape of the definition space, but do not solve the problem of difference in confidence between the center and edges: confidence intervals are larger on the edges than at the center. This problem does not arise in global analyses, but should be considered in local analyses.

Box 5.1. *Monte Carlo simulation and edge effects*

In some cases, it can be shown that the statistical distribution of simulated indices under H_0 is normal, quasi-normal or log-normal. The simulation-based evaluation of the distribution then allows the determination of the characteristics of this distribution (mean and standard deviation) and direct deduction of the *p-value* of the observed situation S_{obs}. In other cases, the *p-value* of the observed situation is determined using its rank with respect to the simulated situations S_k:

$$p - value = \frac{\sum_{k=1}^{N} 1\{|S_k| \geq |S_{obs}|\}}{N}$$

where N is the number of simulations effected and 1 is the indicator function.

The number of simulations required for a good asymptotical approximation of the index distribution under H_0 uses mainly the survey theory. Indeed, the simulations represent a sample from the set of possible situations, and the accuracy of the estimation of the *p value* of the observed situation based on estimated parameters (standard deviation in normal and quasi-normal case, or rank in the general case) depends on the number of simulations performed. It is thus common to conduct between 5,000 and 10,000 simulations to reach a good estimation of the parameter.

Several problems are common in this approach:

– the index does not precisely correspond to the problem to be solved and to the stated hypothesis, or does not offer the possibility to discriminate the alternative hypothesis. This is the most common problem. When a problem is formulated in two dimensions, the elaboration of a numerical index – real-valued function in one dimension – that properly expresses the problem is sometimes difficult: the index may not precisely correspond to the stated spatial hypothesis, or it may produce a poorly discriminating, and therefore powerless, test. Test power corresponds to the probability of choosing Ha when it is actually true;

– even though the index is correct, simulated situations do not correspond to the random situations that must be generated with respect to the problem to be solved. This problem is the most difficult to detect, as there are many traps;

– the power of a test (its capacity to reject the null hypothesis when it is actually false) depends on the spatial configuration of the set D of initial objects. For example, a test for the detection of the cluster character of a subset of points of D is more powerful if D is regular than if D is itself clustered;

– a global result involving several local tests faces the "multiple testing" problem, which is mentioned in Chapter 2.

When the rejection of the null hypothesis H_0 involves several tests, and the rejection of the null hypothesis of a single test is sufficient to reject it, the classic multiple testing problem arises: the risk to mistakenly reject the null hypothesis increases with the number of tests conducted. A classic correction to the multiple testing problem is to divide the risk of error of each test by the number of tests conducted (Bonferroni correction), but this solution is very conservative, as very often the tests conducted are not independent of each other. This is particularly the case in spatial analysis, where there is often a very strong spatial autocorrelation of the p-value of the index. Another solution is to reformulate the global null hypothesis (the rejection of the null hypothesis of a single test is no longer sufficient for the rejection of the global null hypothesis) or to use other techniques, such as the global envelope of ranks [MRK 15].

Box 5.2. *The multiple testing problem*

5.1.2. *Continuous space, point pattern, subsets*

The direct methods used in spatial analysis depend on the type of domain D (support of studied objects, continuous or discrete), and, when studying an attribute, on the type of attribute (qualitative or quantitative – measure or count). As seen in Chapter 2, the following situations are encountered:

– continuous space, analysis of a Boolean variable (presence/absence). For example, analysis of the spatial distribution of a tree species in a forest, the forest being considered as a continuous space. The support D can also be a continuous subset of dimension 1, as in the case of transportation networks in traffic accident and road safety studies, street networks in criminology, or for the spatial analysis of behaviors (for example, the consumption of narcotic drugs in the streets);

– continuous space, analysis of a quantitative variable representing a measure (for example, ore content in the soil), and measured on a discrete subset of D;

– discrete space, analysis of a quantitative variable representing a measure. For example, antibody levels in surveyed individuals located at a place of residence;

– discrete space, analysis of a quantitative variable representing a count (for example, number of patients per geographical unit localized by their centroid), a mean value (for example, mean per geographical unit of the age of puberty for adolescents) or a ratio (for example, proportion per geographical unit of immunodepressed individuals among tuberculosis patients);

– discrete space, analysis of a qualitative or Boolean variable. For example, farms infected with avian influenza among all farms.

Spatial statistics refer to two quite distinct domains: geostatistics, when the domain D is continuous, and the study of the spatial distribution of a point pattern or of their values, when the domain is discrete.

In epidemiology, there are mainly situations in which the spatial definition domain *D* is discrete: either the localization of individuals is limited to places (such as residences, villages, etc.) or the values to be analyzed are already the result of an aggregation into geographical units whose location is synthesized by a centroid (for example, a number of cases per village, an incidence per district). Then, *D* corresponds to a *point pattern*.

Situations where the domain of definition is continuous in dimension 2 are found mainly in the study of environmental risk factors. This refers especially to problems in which a value can be observed at any point on the studied domain, but is only measured at a sample of points. Geostatistics is then used for the study and estimation of a value in any point, depending on the values measured on this sample of points (see Appendix, available online at: www.iste.co.uk/souris/epidemiology.zip).

The situations in which the support D is a continuous domain of dimension 1 refer to health geography studies of events that can only occur in a network (traffic accident and road safety, criminology, etc.).

The objective of *point pattern analysis* is to highlight the characteristics of the spatial distribution of points or of their values. Many questions related to spatial distribution can be formulated, such as: are the points or their values globally distributed in a specific way in the considered domain? Are they clustered or rather dispersed? If the clusters (of points or values) exist, where are they located? Are there any centralities? Are two spatial distributions correlated?

The type of analysis commonly conducted in epidemiology is analysis of the spatial distribution of values (quantitative or qualitative) of points, and not the analysis of the spatial distribution of the points themselves. This is referred to as *marked point pattern*.

When the studied attribute is qualitative and the study focuses on the distribution of a specific value V_0, the studied value allows the definition of a subset D' of the set of points D. Indices using direct topological relationships of adjacency or contiguity between points (for example, frequencies of the nearest neighbors of the same value V_0 or distance to the nearest neighbor of the same value V_0) can then be built [DIX 94, DIX 02]. In case of Boolean attribute (1/0), the term *case/non-case* is often employed.

Spatial analysis of a point pattern is not essentially different from a multivariate analysis, the techniques of which it may use; on the other hand, it has the advantage of implicitly using measures of "natural" distances, stemming from the structure of metric space directly related to the nature of the studied phenomenon (the physical reality of the real world), while "classic" multivariate analyses use a metric built from semantically different variables, which are used as coordinates of the same vector space. This construction involves many hypotheses, and is often difficult to interpret. Several "natural" metrics can be used in spatial analysis. Euclidian distance is obviously the most common.

Spatial analysis of the distribution of events when the spatial support is a network (considered as a continuous subspace of dimension 1) has been the object of specific developments aimed to adapt and limit the classic methods of the continuous case to this type of spatial support.

5.2. Global spatial analyses

The methods used for global analyses are often classified depending on the type of data to be analyzed: on the one hand, the methods used for individual data, and on the other hand, the methods used for data aggregated into geographical units. The localization of aggregated data is actually different: as it will be detailed in the next chapter, it is less connected to the initial phenomenon, since the localization of the initial objects has been lost in the aggregation process; by definition, it is still in the form of zones, which allows the use of certain topological parameters, notably adjacency, that are not directly available in the point data. Nevertheless, the methods presented below can be applied to individual data, represented by a point, as well as to aggregated data related to a zone, and spatially represented by a centroid.

It is always important to specify the subject of the study: events in a continuous space or subspace, events in a discrete space, Boolean or qualitative variable over a set of objects and numeric variable on a set of objects.

The objective of global analyses is to determine the characteristics of the spatial distribution of events or attributes of agents, in order to characterize the underlying process (or processes) and their determining factors. These characteristics refer to: the global synthetic location, the extent and orientation of the cloud of points, the spatial relationships between objects or their values (spatial dependency) and the geometric characteristics of the cloud of points or of values (radial distribution, gradient, shape, etc.). The location of the spatial support of the phenomenon (places where the phenomenon can occur) is not part of the analysis and should not influence the result; for example, if the spatial support is a set of villages, the characteristics of the spatial distribution of the phenomenon in the villages are determined with respect to the spatial distribution of the set of villages, whose characteristics are not part of the study. Nevertheless, as already noted, the spatial distribution of the support may impact the power of statistical tests.

5.2.1. *Geographical location, extent, orientation*

Let us consider a point pattern in a domain of definition D, which can be discrete or continuous. The point pattern may correspond to all the events or to a modality of a qualitative attribute of the points of D. For example, the location of lightning strikes on a territory (continuous case, and no attribute), or the location of "infected" farms in a set of farms (discrete case, Boolean attribute). The methods described below are generally applicable to events, non-aggregated initial data.

The indices use parameters allowing location synthesis and dispersion (similar to the standard deviational ellipses described in Chapter 4).

5.2.1.1. *Absolute geographical location*

The objective is to analyze the absolute global location of the studied phenomenon. The absolute global location of a spatial distribution is characterized by a synthesis object, the mean center. We are testing if the location of the mean center of the observed distribution is significantly far away from the location of the mean centers of simulated distributions. To do this, we use a global mean center, which is the mean center of the mean centers of the simulated distributions, and a standard deviational ellipse whose center is the global mean center, and which contains (100 – α)% of the simulated mean centers (α is the chosen risk of error). The null hypothesis is: "the center of the observed situation does not differ from the global center", and the alternative hypothesis is: "the center of the observed situation differs from the global center". H_0 is rejected if the observed center is not in the global standard deviational ellipse. The test is always one-tailed.

QUESTION.– Globally, can the locations of events be considered as random?

This concerns the continuous case and the absolute location of the set of all events. Each simulation randomly chooses (according to a given model) the location of the events in the domain and calculates the mean center; then, we test the location of the observed mean center with respect to the set of simulated mean centers. The domain may be of dimension 1 for events in a network: then, the simulated events must also be chosen on the network (*network-constrained simulation*).

QUESTION.– Globally, can the location of cases among non-cases be considered as random?

This concerns the discrete case and a Boolean variable that determines the cases in the set of points of the domain. The simulation uses a permutation of values among all points, without changing the location of these points. For example, we study the location of villages infected by a disease in the set of villages (each village being assigned an infected/non-infected Boolean value); the locations of the infected villages are synthesized by a mean center. We compare the location of this center among the centers obtained by simulations by randomly distributing the "infected" values in the set of villages. Each simulation yields a mean center (the mean center of the villages having received the "infected" value). Then, the mean center of these mean centers is calculated. The result is called a global center. This is followed by the calculation of the standard deviational ellipse that contains 95% of the simulated centers (for an α risk of error at 5%). Finally, a test is conducted to see if the center of the actually observed infected villages is within this ellipse (Figure 5.1).

5.2.1.2. *Extent*

QUESTION.– Globally, can the extent of the events be considered as random? Otherwise, is it more dispersed or less dispersed?

QUESTION.– Globally, can the extent of cases among non-cases be considered as random? Otherwise, is it more dispersed or less dispersed?

The extent of a phenomenon can be analyzed by using the same principle as for the study of the absolute location. The extent signifies the manner in which the points of the subset D' are globally dispersed in D. It can be globally estimated by the standard distance of the subset (mean of the distances to the mean center of the subset). Subsets among D are simulated. In the continuous case, this is done by randomly choosing sets of points in D. In the discrete case, simulations involve random permutation of the values of the points. Standard distance of simulated situations is calculated. The null hypothesis is: "the observed standard distance does not differ from the mean of standard distances", and the alternative hypothesis is: "the observed standard distance differs from the mean of standard distances" (double-tailed case). The distribution of standard distances is estimated. H_0 is rejected if the standard distance of the observed situation is longer or shorter by more than $(100 - \alpha/2)\%$ of the distances calculated from simulated situations. This is a double-tailed test. It can be a one-tailed test if we want to separate the "more dispersed" or "less dispersed" hypotheses.

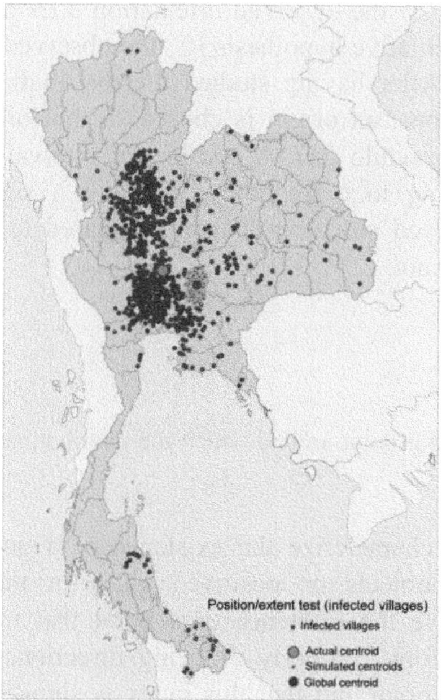

Figure 5.1. *The mean center of observed infected villages (orange) is not within the standard deviation ellipse (blue) of the mean centers of the infected villages by random simulation. This leads to the conclusion that the global location of the phenomenon is not random. For a color version of this figure, see www.iste.co.uk/souris/epidemiology.zip*

The index used here relies on standard distance, which is the mean distance to the mean center. Other indices for testing the dispersion of a value in a cloud of points will be presented further below.

SavGIS	Location and extent tests are available with the menu: Stat → Spatial statistics → Location/Extent
ArcGIS	Location and extent tests are available with the menu: ArcToolbox → Spatial Analyst Tool → Local
R	Location and extent tests are available with the package aspace

Table 5.1. *In practice*

5.2.1.3. *Orientation*

QUESTION.– Globally, can the orientation of cases among non-cases be considered as random? Otherwise, do the cases present a privileged direction?

The null hypothesis is: "the observed orientation does not differ from the mean orientation", and the alternative hypothesis is: "the observed orientation differs from the mean orientation". After having studied the distribution of the orientation of simulated spatial situations, an origin is chosen so that orientation is transformed into an ordered value (modulo 2π). A confidence interval is thus defined for the orientation, corresponding to a hypothesis H_0 and a significance level α. The orientation of the observed situation is then compared to this interval and H_0 is consequently rejected or not.

5.2.2. *Centrality*

QUESTION.– Is there a central point around which the distribution of events or cases among non-cases is radial?

The objective is to characterize the existence of a geometric "center" in the spatial distribution of objects or positive values in the spatial support. The calculation aims to prove the existence of a point that maximizes an index that accounts for the anisotropy (property of being directionally independent) of the cloud with respect to a point and that this index is significantly higher than for a random distribution of points or values. In the discrete case, the spatial distribution of points or values with respect to this center is not necessarily radial: it depends on the configuration of the spatial support.

In epidemiology, the existence of a center allows us to formulate hypothesis on the origin of a proximity-based contagion around a site of infection (Figure 5.2). Centrality is often coupled to an isotropic decreasing (with respect to this central point) of density depending on distance.

A spatial distribution may be generated by several radial distributions (for example, when there is radial diffusion from several sites). It is then difficult to characterize, and the centers are difficult to find relying only on the analysis of the observed spatial distribution. When the time of events is available, the phenomenon can be divided into separate phases (local diffusion from a site followed by a jump at long distance and emergence of a new site of local diffusion).

SavGIS	A centrality test is available with the menu: Stat → Spatial statistics → Centrality
ArcGIS	ArcToolbox → Spatial Statistics Tool → Measuring Geographic Distribution → Central Feature
R	A centrality test is available with the package aspace (CF Calculator)

Table 5.2. *In practice*

Figure 5.2. *The spatial distribution of cholera cases is clearly radial, considering the localization of streets and residences. This visual analysis and the proximity of the center of the spatial distribution to a water supply point allowed John Snow, in 1854 London, to determine the cause of infection [SNO 54]*

5.2.3. *Spatial dependence of values*

These analyses aim to explain the dependence between the values of the objects as a function of their distance: spatial dependence corresponds to the influence of distance on various values. For numeric values, it is analyzed from a statistical point of view in terms of mean and variance in a neighborhood. When the difference between two values does not depend on place, but only on the distance between places, the phenomenon is said to be *stationary*.

> A phenomenon is said to be stationary if the differences between the values of points do not depend on the location of points, but solely on the distance between points. For example, a contagion phenomenon that does not depend on local conditions has the same spatial dependence characteristics regardless of the place of contagion. Global indices, relying on means, *a priori* assume that phenomena are stationary. A general and structural phenomenon or process is naturally assumed to be stationary.

Box 5.3. *Stationarity*

Analyses use various techniques, depending on the type of studied attribute: quantitative attributes involve indices built from distance-weighted values; qualitative attributes involve indices built from the number of objects in a neighborhood and indices using the values of the nearest neighbors.

Most of these indices use a maximal search distance (for example, a maximum influence radius) and are expressed depending on this parameter (*bandwidth*). For numeric attributes, spatial dependence is estimated by an experimental variogram, which includes the statistical information on the observed differences between values depending on the distance between individuals, and which allows the evaluation of the maximal search distance in the calculation of indices. This maximal distance can also be reasoned depending on what is known on the studied phenomenon.

> Semi-variance γ corresponding to a distance nh is defined as the mean of squares of differences between the values z of points in p pairs Pi, Pj so that $nh < d(Pi, Pj) < (n+1)h$:
>
> $$\gamma(nh) = \frac{1}{2p}\sum\{z(P_i) - z(P_j)\}^2$$
>
> After having chosen a unit h, we calculate $\gamma(nh)$ and its standard deviation, n varying from 0 to N so that all the pairs of points are covered. These values are then plotted on a graph, with the distance on the x-axis, and semi-variance and standard deviation on the y-axis.
>
> Variograms often have the shape of a curve starting from an initial non-zero value (kernel, intrinsic variance of the measure), then increasing quasi-linearly (spatial

dependence generally decreases with distance, hence variance increases) and then reaching a level corresponding to the distance above which the phenomenon presents no more spatial dependence (range) (Figure 5.3).

Semi-variance is directly used for calculating the weights in the geostatic techniques for barycentric estimation and interpolation (kriging).

Box 5.4. *Experimental variogram*

SavGIS	The calculation of an experimental variogram is available with the menu: Stat → Spatial statistics → Variogram
ArcGIS	The calculation of an experimental variogram is available with the menu: ArcToolbox → Geostatistical Analyst Tool → Explore Data → Semi-variogram/covariance cloud
QGIS	The calculation of an experimental variogram is available with the menu: Processing Toolbox → SAGA → Geostatistics → variogram cloud
R	The calculation of an experimental variogram is available with the library gstat

Table 5.3. *In practice*

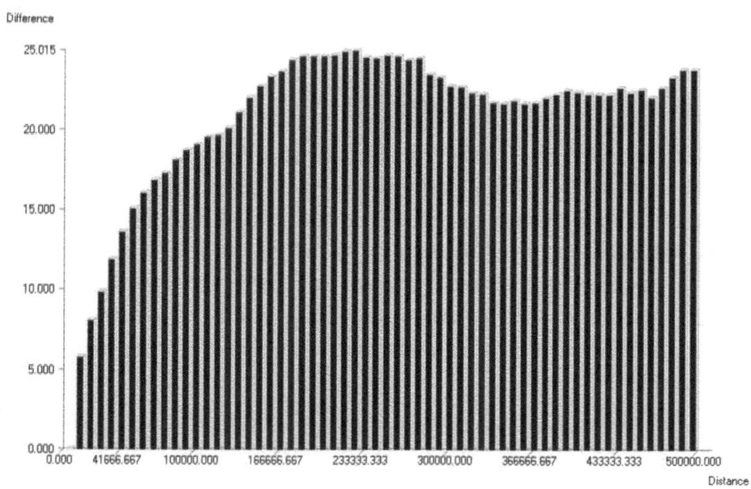

Figure 5.3. *Experimental variogram (mean temperature in Laos in April, calculated by a square grid of 10 × 10 km). The unit h is 10 km. A level is reached at approximately 200 km [SOU 17]*

5.2.3.1. *Clustering, dispersion, uniformity*

QUESTION.– Globally, can the spatial distribution of cases among non-cases be considered as random? Otherwise, is it dispersed, clustered or rather uniform?

Here, the analysis concerns the spatial dependence for a process marked by a qualitative attribute. The analysis focuses on the globally clustered, dispersed, uniform or spatially random character of the location of a subset D’ of points in D. The subset corresponds to a modality of a qualitative variable. For example, the objective of the analysis is the spatial distribution of a tree species in a forest (continuous support), the clustering of traffic accidents (continuous support of dimension 1) or the spatial distribution of “infected” villages in a set of villages (discrete support, qualitative variable, modality “infected”). The objective is to find if the subset presents a spatial distribution that, in terms of distance or vicinity between points, can be distinguished from a random distribution. These methods apply to events, initial point data that are not aggregated into zones.

The null hypothesis is: “the global spatial distribution cannot be distinguished from a random distribution” (CSR – *complete spatial randomness*). The alternative hypothesis for a double-tailed test is: “the global spatial distribution is not random”, and for a one-tailed test: “the spatial distribution is globally clustered” or the opposite “the spatial distribution is globally uniform”. A uniform situation corresponds to the reverse of a clustered situation and to a spatially dispersed situation (no clusters). The null hypothesis corresponds to a global situation: it does not signify that the distribution presents no local clusters. The same is valid for the alternative hypothesis in the uniform case. On the other hand, a clustered alternative situation always presents local clusters.

5.2.3.1.1. *Elaboration of indices based on nearest neighbors*

The underlying idea is simple: the clustering of points can be measured by measuring the distance between neighboring points. Indices are therefore elaborated based on the distance between an object and its nearest neighbors. Several indices are used: the mean of distances to the nearest neighbors (continuous case, events), the mean of the distance to the positive nearest neighbor for positive points (discrete case, Boolean variable) or the frequency of positive points whose nearest neighbor is also positive (discrete case, Boolean variable). Neighbors are defined according to Voronoi [SHA 05].

Voronoi tessellation from a cloud of points D corresponds to a division of the space in as many zones Z_i as there are points P_i in the cloud D. Each zone Z_i is the set of points in space that are closer to point P_i than other points in D. Two points P_i and P_j in D are neighbors in the sense of Voronoi if the zones Z_i and Z_j have a common edge (Figure 5.4). In spatial analysis, Voronoi tessellation is essentially used to determine the neighboring points and to construct spatial weights matrices by neighborhood. The graph that joins all the points that are direct neighbors (order 1) is a triangulation (known as Delaunay triangulation).

Box 5.5. *Voronoi tessellation*

Figure 5.4. *Voronoi tessellation (or Thiessen polygons) corresponds to a sorting in two dimensions; it allows us to find the relative locations between objects (topological relationships). For a color version of this figure see, www.iste.co.uk/souris/epidemiology.zip*

When domain D is continuous, a random situation is simulated by drawing the location of points in D. When domain D is discrete, the distribution of the variable is simulated by random permutation of values between the points of D. The characteristics of the distribution of simulated index values are the basis for rejecting or not the null hypothesis, depending on significance level α.

When the function F is qualitative and a modality *M*0 is chosen, the objects of values M0 define a subset D' among D: $D' = \{Pj \in D/F\ (Pj) = M0\}$, and the problem can be solved from a set-based perspective using the relationships between the points of D' and the points of D, or between the points of D' themselves. Several indices have been proposed in the literature; they rely on the distances between all the objects or between an object and its nearest neighbor(s) (k-NN), particularly the nearest neighbor (NN) [PIE 61, BAR 75, CLI 81, RIP 81, DIG 83, CUZ 90, STO 95, LIE 96].

Isolated points (those that are too distant from their nearest neighbor) can be ignored in the calculation if the distance to their nearest neighbor is estimated to be incompatible with the studied phenomenon. It is thus interesting to ignore in the calculation of the global index the points whose distance to the nearest neighbor exceeds a given limit *d*0. For example, one of the simplest indices is the mean of the distance to the nearest neighbor (DNN) for all the points in D', with an additional condition on the distance to the nearest neighbor:

$$S(d_0) = \frac{1}{n1}\sum_{n1} f(P_i)\ \forall P_i \in \mathrm{D}', with\ f(P_i) = (DNN(P_i, D'))^m$$
$$if\ DNN(P_i, D) < d0 \text{ and } \mathrm{f}(P_i) = 0\ otherwise$$

$DNN(Pi, D')$ is the distance between point Pi of D' and its nearest neighbor in D', and n1 is the number of points in D'. The condition $DNN\ (P_i, D) < d0$ refers to the set D (not only to D'). The value of m is used for weighting the absolute influence of the distance on the index. The distribution of the index is evaluated by simulation, permutating in D the values 0/1 of points, which is equivalent to simulating the subset D' in D.

Indices based on the frequency of points of D' whose nearest neighbor belongs to D' can also be used:

$F = Card\{P, P \in D' and\ NN(P) \in D'\}/Card(D')$, where Card(D) designates the number of objects in D.

The test can be double-tailed or one-tailed. In the one-tailed case, there are two distinct situations: the observed value is significantly lower than the simulated value (the global distribution is rather clustered), and on the other hand, the observed value is significantly higher than the simulated value (the global distribution is rather uniform).

A uniform spatial distribution is a very particular case of dispersed distribution (it is the least probable case of dispersion). If domain D is continuous, it corresponds to the absolute distribution in space; the most uniform case is a triangular distribution, in which all the distances between neighbors are equal. If domain D is discrete, then the spatial distribution of D must be considered. If only the uniformity of a distribution is to be tested (and not the clustered or dispersed character), then an index that accounts for this property must be formulated. For example, the standard deviation of distances between points can be chosen.

If domain D is discrete, the power of these tests (their capacity to detect the alternative situations and reject the null hypothesis) depends on the spatial distribution of the points in D themselves: the more clustered the points in D, the less powerful the test allowing the detection of clustering in the subset D'.

SavGIS	The clustering tests using neighbors are available with the menu: Stat → Spatial statistics → Vicinity/Distance

Table 5.4. *In practice*

5.2.3.1.2. Elaboration of indices based on the number of events in a disk

The analysis of clouds of points also uses statistics built on the number of events or positive values in a disk of radius $d0$. The idea is to calculate a global index from the set of pairs of points whose distance is lower than a given value $d0$, and not only between pairs of neighboring points. Ripley's function K ($d0$) [RIP 81] and its normalized version H are expressed by:

$$K(d0) = \frac{1}{\lambda n}\sum_{i=1}^{n} N_i(d0) \; and \; H(d0) = \sqrt{\frac{K(d0)}{\pi}} - d0$$

where N_i ($d0$) is the number of events (or positive values in the discrete case) in a circle of radius $d0$ centered on the point i, n is the total number of observed events (or the number of points in the discrete case) and λ is the estimation of the global density of events (or the global percentage of positive points in the discrete case).

Ripley's functions are mainly used for the study of spatial processes in the continuous case. They have the advantage of allowing a description of the characteristics of processes at various scales. Thus $d0$ can be varied in order to find the value that maximizes H. Similar to the indices built from the nearest neighbors, the index H can be used to test a null hypothesis of random distribution against an alternative hypothesis of clustered or uniform distribution. As usual, the distribution under null hypothesis is obtained by simulation.

Tango has also developed global autocorrelation indices (EET and MEET, for *excess event test* and *maximized excess events test*, respectively) based on the number of events clustered into zones [TAN 95, PFE 08]. They are built as a mean of excess events (difference or ratio between the observed rate and the expected rate) weighted as a function of the distance between the two objects of the pair. They use the SMR (see Chapter 6) rather than the direct value of objects (rates or numbers of events). These indices have the disadvantage of requiring the *a priori* definition of the cluster zones (defined most often as disks of given radius).

5.2.3.2. *Autocorrelation: elaboration of indices based on pairs of points*

QUESTION.– Globally, is there a relationship between the spatial distribution of the value of objects and the proximity of objects? If yes, what is this relationship? Is the residue of a regression spatially random?

When the studied variable is quantitative, the set-based and topological methods used in the previous section in the case of counts or of a qualitative variable are not applicable. An index called spatial autocorrelation is then used to measure the spatial interdependence of values [CLI 69]. Domain D is continuous or discrete, and the function F is known only on a finite number of points that correspond to the events or to the places where the variable has been collected or measured.

Spatial autocorrelation is a statistical concept used to express the correlation between the values of objects depending on the metric or topological relationships between objects. It aims to show that the spatial distribution of the values of a variable is not random and to account for the spatial interdependence of values. Moreover, this spatial dependence is considered as the basis of geography ("everything is related to everything else, but nearby things are more related than distant things" [TOB 70]). On the other hand, knowing that the values of a localized variable are spatially dependent allows us to use this dependence when the objective is to estimate an unknown value at a point, or to find a statistical model to explain a variable by a set of other explanatory variables. A component linked to autocorrelation is then added to the model, when it is not explained by the other explanatory variables and when the residue shows a non-random spatial distribution, in order to eliminate – without explaining it – this residual spatial component.

The measurement of this spatial autocorrelation uses a numerical index that allows the expression of the spatial autocorrelation into a numerical value: a spatial autocorrelation index is a statistical measure of the correlations between the values of the localized objects using metric or topological relationships between these objects. An autocorrelation index measures if globally the nearby values have a stronger tendency to be similar than the distant values. It aims to account for the spatial dependence of values and to show that the geographical distribution of these values is not random. The autocorrelation indices use the differences in value and the differences in location of the pairs of points. Their objective is to measure the spatial autocorrelation of function F over its domain of definition D.

The most efficient way to show the existence of a spatial association is to test the statistical significance of the spatial autocorrelation. The autocorrelation indices allow us to study the clustering or spatial dispersion (global or local) of the values and identify the role of contiguity or distance in spatial interactions. A spatial autocorrelation index may be a global mean measure, or it may refer only to the vicinity of a place.

Many spatial autocorrelation indices have been developed. Most indices are built on pairs of points and derive from the index described by Mantel [MAN 67] for the study of clusters of events in time. The Mantel index quantifies the relationship between proximity in time, measured by $|T_i - T_j|$, and proximity in space, measured by a number w_{ij} (based on distance or adjacency), for all the pairs of points P_i, P_j:

$$M_1 = \sum_{i,j,i<j} w_{ij} \times |T_i - T_j|$$

Statistical significance of the index is measured by a Monte Carlo simulation.

Several spatial autocorrelation indices use only the difference in values of the points of the pair, limiting the calculation to pairs whose distance is below a given threshold $d0$ (for each point, this amounts to studying the relationships with the other events located within a disk of radius $d0$). This is the case for *pair correlation* functions and *mark correlation functions*, which are extensions of Ripley's K-functions. The threshold $d0$ is chosen as a function of the maximum distance of supposed influence of one event on another, and depends on the studied phenomenon. For example, a function of correlation between pairs of points can use the product of values of points of the pair [STO 00]:

$$K(d0) = (\overline{X_i X_j})/(\bar{X}\,\bar{X})$$

where $\bar{X}$ is the mean of X for all i, $1 \leq i \leq n$, and $\overline{X_i X_j}$ is the mean of $X_i X_j$ for all $i, j, 1 \leq i, j \leq n, i \neq j \,/\, d(Pi, Pj) < d0$

Similar to the Mantel index, most indices use the value of the pair weighted by a spatial weight, built from a topological or metric parameter depending on the pair (for example, the inverse of the distance between the two points of the pair). The geometric or topological relationship between the two points of a pair must express the spatial dependence and be transformed into a numeric value in order to build a numeric index. There are several possible options for this purpose, depending on how the spatial dependence should be expressed:

– by considering the vicinity relationship. Direct neighborhood between the two points of the pair (according to Voronoi) can be used, assigning 1 if the two points are neighbors, and 0 otherwise. When contiguity or adjacency relationships are considered, the length of the common edge between the two objects (either Voronoi tessellation for a point pattern or the length of the border for adjacent zones);

– by considering the distance between points (or centroids in the case of zones), notably to express the relationship between variance of value and distance between points. A function of the inverse of distance is then used. This function can be linear (for example, $(d0 - d)/d0$), quadratic, Gaussian or sigmoid). It is often limited to a maximal distance, beyond which the value is 0, signifying an absence of spatial dependence beyond this distance. The maximal distance can be fixed for all pairs or can be related to a characteristic linked to density (for example, function of the distance to n-nearest neighbor of one of the points of the pair). It can be globally estimated, over the entire domain, by the range of the variogram corresponding to the situation to be analyzed;

– by considering the mean difference between values depending on distance, over the entire domain (given by a variogram).

Most of these indices are (weighted) means; therefore, they implicitly assume that the phenomenon is *stationary*.

Moran's index is thus defined as the mean of the products of the normalized values of the pairs of points, weighted by a "spatial weight" depending on the contiguity or distance between the two points of the pair:

$$I_{Moran} = \frac{1}{S} \sum_{i,j,i<j} w_{ij} (\frac{Xi - m}{\sigma}) (\frac{Xj - m}{\sigma})$$

where m is the mean of X_i in D, σ is the standard deviation of X_i in D, w_{ij} is the spatial weight of the pair of points (P_i, P_j) and S is the sum of the spatial weights. The pairs (P_i, P_j) are not considered.

Moran's index (or Moran's I) corresponds to a classic correlation index between two variables (Pearson), but with only one variable, the calculation being extended to the values of neighbors and featuring a weighting that depends on the spatial relationships between neighbors (contiguity, adjacency or distance). It uses a multiplicative model. It is the most frequently used index in practice.

Spatial weight is essential in the definition of weighted indices. Contiguity has several meanings: neighborhood (in the sense of Voronoi), existence of a common border (adjacency, in the case of already constituted zones), existence of a common node (still in the case of zones), k-order contiguity (consideration of neighbors of neighbors, up to order k) and distance below a given radius. In Moran's initial definition [MOR 48], the weight w_{ij} of the pair (P_i, P_j) is 1 if the two objects are adjacent and 0 otherwise. The length of the common border between the two zones can also be considered.

Direct methods favor the consideration of the real distance between points. Spatial weight is generally a function of the inverse of distance. A distance d_0 (bandwidth) is commonly defined, which allows parameterizing the weight decrease depending on distance. In this way, non-neighboring values can be considered, if they are assumed interrelated for the studied phenomenon.

For example, the following can be written:

$$w_{ij} = \max\left(0\,, 1 - \frac{d(P_i, P_j)^k}{d0^k}\right)$$

(assigning 0 to pairs of points with a distance above d_0. Power k is used for the parameterization of the decrease)

or:

$$w_{ij} = \exp\left(-\frac{d(P_i, P_j)^2}{d0^2}\right)$$

When anisotropy is assumed, spatial autocorrelation indices can be limited to a given direction, to test the autocorrelation corresponding solely to this direction, and increase the relevance of the index. Spatial weight will then depend on the direction of the vector defined by the pair of points.

The value of $d0$ depends on the studied phenomenon. It generally corresponds to the maximal distance beyond which there is no more spatial dependence. It can be estimated by means of a variogram.

The distance function d used is not necessarily the Euclidian distance. Depending on the studied phenomenon and on the mode of interaction between objects, which is the assumed cause of autocorrelation, it is sometimes more realistic to consider a pathway distance in a network or to consider a pathway time rather than a line distance. For example, Euclidian distance will rather be used for physical phenomena (temperature, humidity, pollution, etc.)

Box 5.6. *Spatial weight*

SavGIS	The calculation of a spatial weight matrix is available with the menu: Calculate→ Spatial weight matrix The spatial weight matrix can be exported and used in other software programs.
ArcGIS	The calculation of a spatial weight matrix is available with the menu: ArcToolbox → Spatial Statistics Tools → Modeling Spatial relationships → Generate Spatial Weight Matrix
QGIS	The calculation of an experimental variogram is available with the menu: Processing Toolbox → SAGA → Geostatistics → variogram cloud
R	The calculation of a spatial weight matrix is available in the package spdep

Table 5.5. *In practice*

The expected value of Moran's I is $\frac{-1}{(n-1)}$, where n is the total number of points in D. Its absolute value increases with the autocorrelation: a value close to 0 indicates an absence of spatial autocorrelation, a positive value indicates positive autocorrelation (close values have a tendency to be similar) and a negative value indicates negative autocorrelation (close values have a tendency to be distant). The statistical distribution of the index under the null hypothesis (random spatial distribution) must be used to verify if the observed situation presents a significant spatial autocorrelation. The statistical distribution depends on the spatial distribution of D (shape and edges) and the number of points. It results from the Monte Carlo simulation, by random permutation of the values of points, without modifying the number and the location of the points. It is used to calculate the p-value of the observed situation and deduce its significance.

Another index of spatial autocorrelation, Geary's index, is built on an additive model, rather than on a multiplicative model: it is defined as the mean of squared differences of the normalized values of the pairs of points, weighted depending on contiguity or distance between the two points of the pair:

$$I_{Geary} = \frac{1}{S} \sum_{i,j,i<j} w_{ij} \left(\frac{Xi - m}{\sigma} - \frac{Xj - m}{\sigma}\right)^2$$

Geary's index values range between 0 and 2, with 0 indicating perfect spatial autocorrelation, and 2 perfect negative autocorrelation. Its variance is also approximated by a Monte Carlo simulation.

Other spatial autocorrelation indices are built according to similar principles, but without using normalized variables and are therefore less frequently used: Black Black Seal, Black White Join, Knox [SHA 05].

Spatial autocorrelation indices are widely used, but they face several challenges:

– the differences in density of points are not considered, and this is annoying when spatial weight involves contiguity, rather than distance: the influence of isolated points is too strong and diminishes the power of the test. In fact, the use of contiguity without considering the distance in the spatial weight poses a problem when the studied phenomenon depends on distance as much as on the notion of neighborhood;

– this problem is also found with the population densities exposed when using counts. Indeed, high-density areas tend to be smaller than low-density areas, and taking solely into account adjacent neighbors, which may generate a masking effect by hiding very close areas;

– when studying autocorrelation of rates, the problem of differences in accuracy that has already been signaled for mapping is also present in the calculation of autocorrelation. EBE-based rate adjustment may therefore be useful before the calculation of autocorrelation indices;

– indices rely on the hypothesis of stationarity, but this hypothesis is rarely verified in epidemiology, particularly for infectious diseases, for which the intensity of infection is rarely evenly distributed in space;

– a non-stationary distribution may present positive autocorrelation, while the values are proved to be uncorrelated. The spatial distribution of D influences the calculation of autocorrelation. More generally, the links between spatial heterogeneity and autocorrelation are complex, as the presence of either one of them can alter the test properties of the other;

– in most cases, the number of pairs of points increases with the distance between the points, reaches a maximum and then decreases again until reaching the edges of the domain (Figure 5.5). When the weights use distance, weighting by this parameter is also required; otherwise, the influence of pairs with medium distances is significantly increased in the global mean;

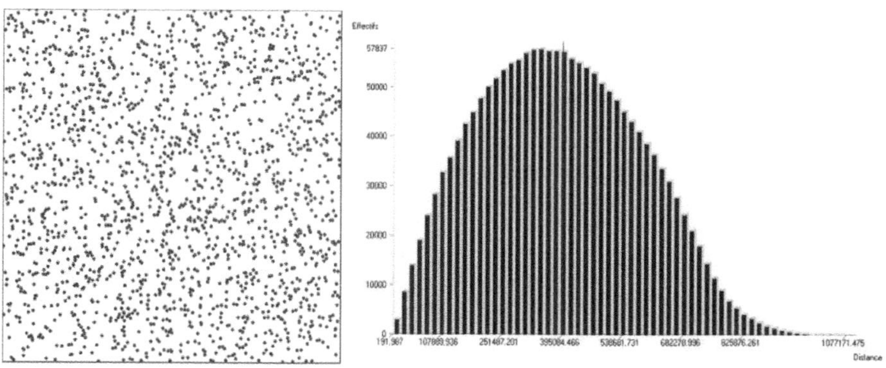

Figure 5.5. *In a finite set of points (of random spatial distribution here), the number of pairs of points increases with distance, reaches a maximum and then decreases*

– specifying a maximum distance *d*0 is difficult to justify, unless subjective considerations are at play (range of the phenomenon: assumed maximum distance of influence of one value on another value). It is recommended to use a variogram or to analyze the influence of *d*0 on the result of the autocorrelation index. It is also possible to directly use the variogram for the definition of spatial weights, when it shows a spatial dependence, and if the data set used is estimated to be representative for the spatial phenomenon;

– finally, there is no direct relationship between the two most commonly used indices (Moran and Geary).

Spatial variability of the index can itself be studied, as it offers an indication of the stationarity of the phenomenon. For each point P_i, the sum of index elements related to point P_i can be calculated: for example, for Moran's index,

$$I_{Moran}(P_i) = \frac{1}{Si}\frac{(X_i - m)}{\sigma}\sum_{j,j\neq i} w_{ij}\ (\frac{Xj - m}{\sigma})$$

and Moran's I variance $I_{Moran}(P_i)$ can be calculated. If variability is high, it may be useful to visualize and analyze the spatial distribution of this variability: the local indices for spatial association will be presented further below. These indices use the same principles as the global indices, but they are only calculated in the neighborhood of a point. They allow us to avoid the problems of global stationarity, but cannot be used for the global characterization of autocorrelation.

An extension of spatial autocorrelation indices to spatio-temporal data will be presented in Chapter 7.

SavGIS	Spatial autocorrelation tests are available with the menu: Stat → Spatial statistics → Spatial autocorrelation indices
ArcGIS	Spatial autocorrelation tests are available with the menu: ArcToolbox → Spatial Statistics Tools → Analyzing Patterns → Spatial autocorrelation
QGIS	Spatial autocorrelation tests are available with the menu: Processing Toolbox → SAGA → Geostatistics → Global Moran's I
R	Package spdep
GeoDa	Temporal animations can be created with the menu: Space → **Univariate Moran's I**

Table 5.6. *In practice*

5.2.4. *Bivariate spatial analysis*

QUESTION. Is the spatial distribution of a phenomenon close or distant from the spatial distribution of another phenomenon? Are the spatial distributions of two phenomena correlated?

Bivariate spatial analysis concerns the spatial relationships between two phenomena. The objective is to use a statistical method to highlight the existence of a global spatial relationship between two processes based on their observed events, for example, the joint distribution of two pathologies in a population, two antibody levels and the same pathology at two given instants. Indeed, two phenomena may have the same mechanisms and the same factors. There are many situations in which a direct proximity relationship is suspected between the events of two phenomena, particularly when the same environmental risk factor is suspected for both phenomena, for example, the same vector, the same reservoir or the same technological risk. In such situations, there is an expectation to highlight that, globally, an event of one of the phenomena is "always quite" close to an event of the second phenomena (or, conversely, "quite distant"). This comes down to highlighting a correlation between two (qualitative or quantitative) variables in pairs of individuals, created from a set of individuals and a spatial relationship between these individuals and weighted by this spatial relationship. The analysis differs from a classic multivariate statistical analysis, which analyzes the individuals only based on their attributes ($y = f(x1, \dots, xn)$). In this case, the analysis concerns only the values of the pairs of individuals and involves their proximity relationships.

The search for spatial relationships between several phenomena arises in many fields, particularly ecology, geology, information sciences (data mining), criminology, land-use planning and economics [WAR 85, GAT 96, SHE 01, PER 06]. Knowledge of one phenomenon can help in decoding and preventing another phenomenon. Many spatial analysis methods have been described in the statistical literature, and this domain has been subjected to in-depth studies [BAR 75, CLI 81, RIP 81, DIG 83, SHA 05, PFE 08]. Nevertheless, the literature almost always presents the multivariate spatial analysis in relation to point processes (mainly in forestry or mining) where both locations and point marks are subject to change [PEN 92]. In epidemiology and health geography, the context is that of marked point processes for which the support location is formed by two sets of points whose location is fixed and is not part of the analysis. This case requires no particular hypothesis on the stationarity or isotropy of the spatial distribution of points, unlike the studies on point processes with continuous support in space.

5.2.4.1. *General case for quantitative variables*

The general case corresponds to giving a numeric value to each of the points of two sets of points *U* and *V*, formally corresponding to two fields of random variables $f: U \to \mathbb{R}$ and $g: V \to \mathbb{R}$ whose realization is known. An index LS [LEE 01][SOU 11] is then defined:

$$LS = \frac{1}{S}\sum_{i,j} w_{ij}\left(\frac{f(Pi) - m1}{\sigma 1}\right)\left(\frac{g(Pj) - m2}{\sigma 2}\right)$$

where m1 is the mean of f on U, σ1 is the standard deviation of f on U, m2 is the mean of g on V, σ2 is the standard deviation of g on V, w_{ij} is the spatial weight of the pair (P_i, P_j) and S is the sum of spatial weights. The expressions $\frac{f(Pi)- m1}{\sigma 1}$ and $\frac{g(Pj)- m2}{\sigma 2}$ correspond to centered-reduced values of the functions *f* and *g*. The spatial weight can be defined in various ways, similar to the univariate case, but here it concerns the pairs of points (P, Q), where P belongs to U and Q belongs to V. Rather than using adjacency, distance will be used, with the same remarks as for Moran's index: similar to the univariate case, the index calculation needs to consider the distribution of distances of pairs (P,Q) to avoid putting at disadvantage in the calculation of the mean the pairs of small distances, less numerous.

The distribution of index under null hypothesis is obtained by simulation by permutation of the values of f and g within the sets U and V. Since the objective is to highlight the spatial correlation between the two phenomena, and not the spatial characteristics of each component, the influence of these characteristics must be minimized, and especially the influence of the spatial autocorrelation of each component in the evaluation of the statistical distribution under null hypothesis. It is therefore useful to select the simulated situations depending on their spatial autocorrelation. Only the simulated situations that meet a criterion of spatial autocorrelation will be retained for estimating the distribution of the index. It is also possible to simulate by permutation only the values of one of the two sets, keeping the values of the other set fixed. The global test is then decomposed into two distinct tests.

5.2.4.2. *Particular case for quantitative values*

The cases when U and V are identical or when their intersection is non-null are very common in practice and must be presented in detail. The objects of $U \cap V$ have values at the same time for *f* and for *g*, and these values can be correlated within the same object, and therefore without any spatial relationship being involved in this correlation. To exclude these "intrinsic" correlations and consider only the spatial correlations between various objects, the index LS* is used:

$$LS^* = \frac{1}{S} \sum_{i,j,Pi \neq Pj} w_{ij} \left(\frac{f(Pi) - m1}{\sigma 1}\right)\left(\frac{g(Pj) - m2}{\sigma 2}\right)$$

The index LS* is at the same time an extension of the classic correlation index to the values of "neighboring" objects (according to the definition of spatial weight considered), and an extension of Moran's autocorrelation index to the spatial correlation of two functions.

In the simulation process used for approximating the distribution of this index under null hypothesis, the permutation of values must not destroy the intrinsic correlation: it must preserve the values of f and g in the pairs (P_i, P_i).

5.2.4.3. *General case for Boolean qualitative variables*

When the functions f and g are Boolean, a set-based approach similar to the univariate case can be used, each function defining a subset (U' of U and V' of V) corresponding to the points whose value is 1. A first visual approach can be used, comparing the standard deviational ellipses of two subsets, but this approach has no statistical validity.

Bivariate statistical analysis uses the spatial relationships between individuals of two subsets U' and V'. The significance of the difference of some geometric elements of the distributions can be tested (distance of the centers of two distributions; differences of dispersions; differences of global orientations), but the most complete test uses an index calculated from distances to the nearest neighbors (from a point of U' to its closest neighbor(s) in V'). The distribution of the index under null hypothesis is obtained by simulation of U' or V', by permutation of values 1 within U and V.

Similar to the case of quantitative variables, to avoid the introduction of a confounding factor due to intrinsic characteristics of spatial subsets (that is, clustered or regular), the global test is divided into two distinct tests, by maintaining a fixed subset and randomly choosing (under condition on spatial characteristics) the other subset for estimating the distribution of the index under the null hypothesis.

The simplest index, function of a maximum distance *d*0, is built from the mean of distances DNN(P, U'), for all the points P of V'.

$$\text{ST}(d0)= \frac{1}{n2}\sum_{n2} f(Pi)\ \forall Pi \in V', with\ f(Pi) = (DNN(Pi,U'))^m$$
$$if\ DNN(Pi,U') < d0\ and\ f(Pi) = 0\ otherwise$$

where $DNN\ (Pi, U')$ designates the distance between point P_i of V' and its nearest neighbor in U', n2 is the number of points of V' and *d*0 is the exclusion distance, beyond which there is no reason to search for interactions between the two phenomena. The value of m (1/2,1 or 2) is used to adjust the absolute influence of distance.

Similar to the numeric case, when the two sets U and V are identical, $U' \cap V'$ can be non-null (the same object P_i can be positive for both phenomena), but these double values 1-1 in the same P_i can have no relationship with the localization and no intrinsic correlation should be considered as spatial. The simulations used for estimating the characteristics of the distribution of the index must maintain the global intrinsic correlation (same number of values 1-1).

The interpretation of this index is obvious. If $ST(d0)$ is significantly smaller than expected under randomization, the mean of distances from a point of U' to its nearest neighbor in V' is smaller than under null hypothesis: the positive values of V are globally nearer to the positive values of U than in the simulation. If $ST(d0)$ is significantly larger than expected under randomization, the positive values of the two sets generally show a repulsion globally.

5.2.4.4. *Particular case: source points*

When one of the sets of events is fixed, and the analysis only focuses on the location of the events of the second set of events in relation to the first set of events (this is often the case when searching for a cause and effect relationship between the two phenomena), the bivariate analysis is similar to a search for clusters of events of the second set around the points of the first set, considered as source points.

The analyses around a predefined source point (or line) (*focused cluster analysis*) are similar to the search of clusters and of excess risk around a place and will be described in the next chapter.

When instead of one source point the analysis focuses on a set of points, it corresponds to a bivariate global analysis in which spatial variability concerns only one of the two phenomena, the other set being fixed. The methods used for only one source point can be extended to a set of points to address the global question.

5.2.4.5. *Example*

In the following example, the analysis concerns a georeferenced set of serum from patients from Timone Hospital in Marseille (472 serums, located in the city of Marseille, randomly selected from the hospital serum bank, parasitology unit) [SOU 11]. Two Boolean variables (with two modalities: yes/no) are available for each serum: the presence of IgG Leishmania, and the presence of IgG Toscana. The objective is to estimate the spatial correlation between these two variables, independently of the type of initial spatial distribution, in order to highlight the possible implication of the same vector in the diffusion of infectious agents related to the presence of these antibodies and to determine the spatial footprint (Figure 5.6). The minimum mean distance between the points of the sample is 245 meters.

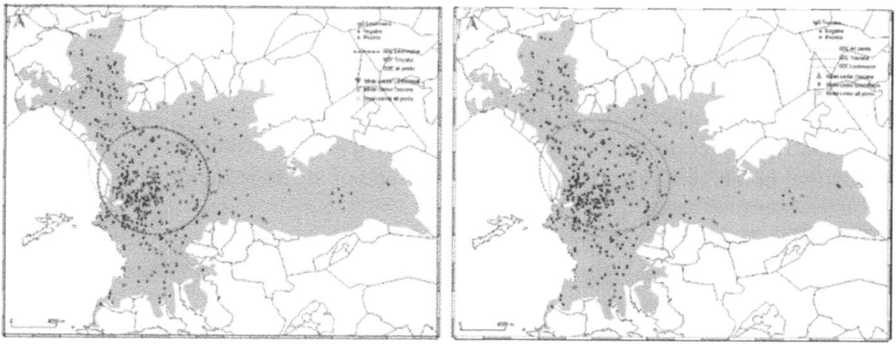

Figure 5.6. *Distribution of IgG Leishmania (left) and IgG Toscana (right) in the study zone. The orange points represent positive serums. The standard deviational ellipses give a summary of the extent and mean location of each of the point patterns. For a color version of this figure, see www.iste.co.uk/souris/epidemiology.zip*

The classic statistical analysis shows a significant relationship between IgG Leishmania+ and IgG Toscana+ (χ^2 = 38.3) among the serums (intrinsic correlation).

The standard deviational ellipses offer an indication of the absolute spatial distribution of points and allow the visual comparison of various sets (blue for Toscana+, purple for Leishmania+). The index LS* does not allow the rejection of the null hypothesis: with our sample, the correlation between IgG Leishmania+ and igG Toscana+ is limited to the intrinsic correlation, and the distance between patients shows no relationship with the co-infection phenomenon. The intrinsic correlation must be explained: either the distance of influence of a possible common environmental factor is below the mean distance of the nearest neighbor of our sample (245 m) or double infections have no direct relationship with the place of residence of the patients.

SavGIS	Bivariate analysis tests for numeric variables are available with the menu: Stat → Spatial statistics → Bivariate correlation index Bivariate analysis tests for Boolean variables are available with: Stat → Spatial statistics → Colocalization

Table 5.7. *In practice*

5.2.5. *Global pattern of the phenomenon and search for a geometric model*

Besides simple global characteristics of the spatial distribution (centrality, autocorrelation, dispersion, continuity of values), it is worth finding a global mathematical function that models the observed spatial distribution (of the location of positive points or values, depending on the type of attribute under study).

QUESTION.– Is there any mathematical function (preferably a simple one) that models the spatial distribution of the phenomenon? Is there a process that generates the spatial distribution of the phenomenon?

The objective is to:

– approximate the observed spatial distribution with a simple mathematical pattern (such as a linear or quadratic surface or a Fourier series), based on the geometric characteristics established from the previous results (centrality, dispersion, direction, anisotropy, decreasing from a center, cycles, periodicities, depending on a direction, etc.);

– bring in hypotheses on the value generation processes (for example, Poisson or Cox process modeling for continuous case or Bernoulli or negative binomial process for discrete case, etc.). Hypotheses concern the assumed characteristics of the phenomenon, some of which can be determined by preliminary analyses (place of index cases, centrality, dispersion, etc.) and others by *a priori* knowledge on the phenomenon (contagion process; propagation process).

SavGIS	The simulation of contagious spatial processes is available with the menu: Calculate → Spatio-temporal calculation → Contagion

Table 5.8. *In practice*

5.3. Local spatial analyses

In the previous sections, we have described the indices of global spatial autocorrelation, which are elaborated as means between the values of objects and the values of their neighbors on the whole studied territory, and which implicitly assume that the phenomenon is stationary (the process that generates the spatial distribution is identical everywhere) since the objective is to characterize it by a mean. It is nevertheless possible, even under this hypothesis of stationarity, to take interest in the spatial variation of autocorrelation, particularly when these variations are significant. Significant local spatial associations between values may exist, but they yield an insignificant global autocorrelation index. These local spatial associations can thus be hidden in a global analysis.

The objective of the local spatial analysis is to detect particular patterns or distinctive places, such as local concentrations of high values, clusters (of objects or values), local centers, attraction places.

Let us recall that during an explanatory analysis, the objective of all spatial analyses, either global or local, is to decode and understand the global processes that have induced the spatial distribution of the health phenomenon. If the process that generates the distribution is assumed global, there is a danger to use local analyses in order to characterize places, without considering the characteristics of this global process. Nevertheless, local studies may allow the detection of potential risk factors (mobile or immobile), such as the presence of vectors or a concentration of socio-economic vulnerabilities. Local analyses also allow direct impacts in terms of public health, reversing the cause-and-effect relationship, when the mechanisms of a phenomenon are known: if a phenomenon is known to involve a clustered spatial distribution of cases (for example, infectious diseases), the real-time detection of these local clusters allows us to activate early warning systems and follow up the phenomenon.

The next section focuses on the study of local indicators of spatial association (LISA) and the detection of distinctive places by spatial scan. Analyses of excess risk around a point or a predefined source line (focused cluster analysis) are addressed in the next chapter.

5.3.1. *Local indicators of spatial association (LISA)*

QUESTION.– Are there places for which the spatial distribution of the value of objects has a significant relationship with the proximity of the objects? Where are the high-value clusters? Where are the low-value clusters?

Even though the phenomenon does not present global autocorrelation, and especially if the spatial variability of local elements in the global index of autocorrelation is important, it is interesting to study local spatial associations. This is all the more true as the domain is wide and presents marked regional differences.

The local indicators of spatial association (LISA) measure the "concentration" of numeric values measured on a set of points or calculated by spatial aggregation. They rely on the same principles as the autocorrelation indices, but the calculation is made for each point P_i using only the neighbors of this point. From the observed situation, an index is calculated for each point P_i, and a permutation-based simulation is made for each point P_i, by permutating the values of neighbors to approximate the null distribution of the index and deduce the probability p of the observed value. When the value of p does not allow the rejection of the null hypothesis, point P_i is not considered to have particular character. When the value of p is below the significance level α, several cases are possible: point P_i has a strong value and is surrounded by strong values ("hot spot"), point P_i has a weak value and is surrounded by weak values ("cold spot") and point P_i is surrounded by values that are opposite to it ("distinctive spot"). The graphic representation of these situations is known as a *Moran scatter plot* [ANS 96] (Figure 5.7).

The cloud of points, or Moran scatter plot, allows us to visualize the relationships between the value of a point P and the mean of values of the neighboring points. The x-axis represents the standardized values V, and the y-axis the standardized means of neighboring values NV, and each point of the graph corresponds to a pair (V, NV), corresponding itself to a point that is statistically significant for the local index. The plot is centered on the value (0,0) corresponding to the mean of the observed values. The right upper quadrant corresponds to hot spots (strong values surrounded by strong values), and the left lower quadrant to cold points (weak values surrounded by weak values). The right lower quadrant and the left higher quadrant contain the points with negative spatial association.

Box 5.7. *Moran scatter plot*

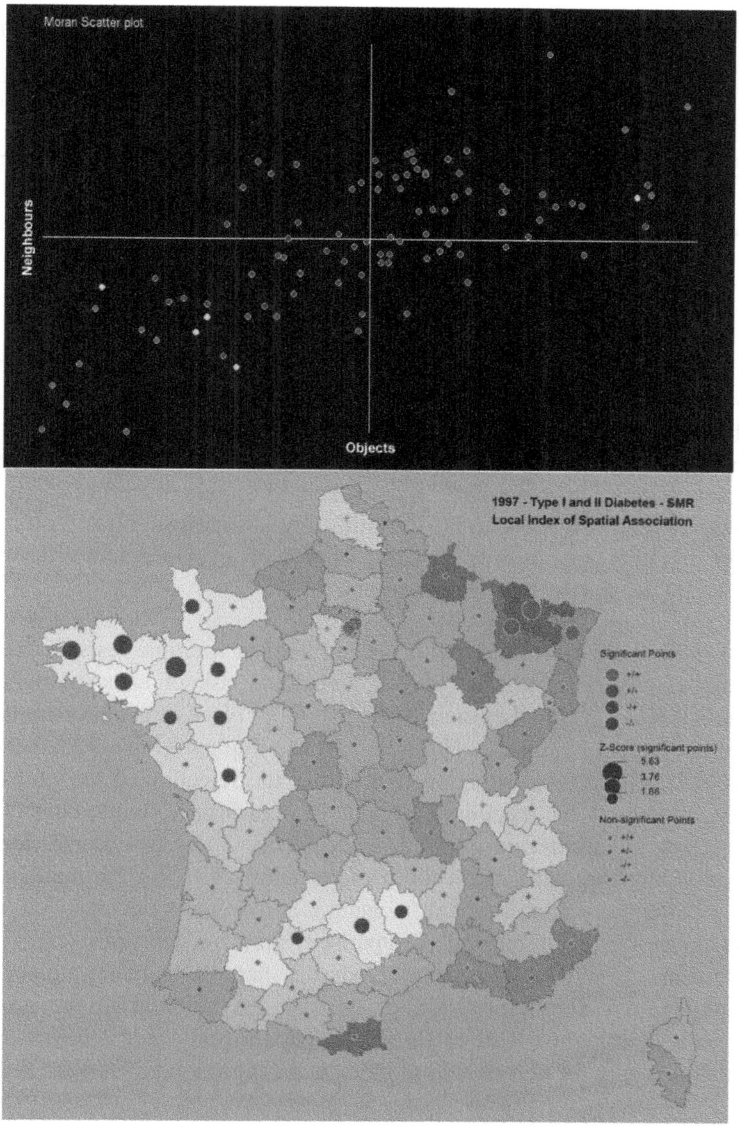

Figure 5.7. *Moran scatter plot (diabetes in France, 1997). The points in yellow and red on the plot correspond to statistically significant situations (risk α at 5% in yellow and at 1% in red); the other points (in purple) are not statistically significant. The points in blue on the map are cold spots (weak values surrounded by weak values), and the points in red are hot spots (strong values surrounded by strong values). The points not statistically significant are represented by a rhomboid. For a color version of this figure, see www.iste.co.uk/souris/epidemiology.zip*

SavGIS	The calculation of *Moran Scatter plot* is available with the menu: Stat → Spatial statistics → Moran Scatter plot
ArcGIS	The calculation of *Moran Scatter plot* is available with the menu: View → Graphs → Create Scatter plot Matrix Graph
R	Spdata package Spdata package (Moran_plot)

Table 5.9. *In practice*

The crucial question is obviously that of the evaluation of the null distribution at each point P_i, since this evaluation allows the calculation of the probability p of the value observed at point P_i: the neighbors to be considered for this evaluation should be defined. It makes no sense to consider all the points of domain D, since we are looking for a local characteristic. The evaluation should therefore be limited to points that are relatively close to P_i. The difficulty lies in this choice, as it requires a definition of either an absolute distance of influence or a relative distance that depends on the local density of points around P_i.

Local indices are confronted with the problem of multiple tests. Indeed, when conducting a local test for each object, the global risk of error is higher if the question relates to the global existence of a particular place. It is important to specify the "implicit" global hypotheses H_0 and Ha. If the implicit global hypothesis Ha is: "there is at least one local situation that allows the rejection of the global H_0", a correction must be used, reducing the value of significance, while considering the fact that close situations are not independent (the *Bonferroni* correction, which involves dividing the value of the limit of rejection of H_0 by the number of tests effected, is often very conservative). If using a correction is not intended, then any global analysis should be banned (comparison, gradient, tendency, etc.) and only local questions should be considered ("what is happening in that place?"), and should avoid a visual interpretation of the map. A solution to the problem of multiple tests is brought by the methods that maximize an index, before searching the significance only for objects whose index is the highest (see section 5.3.2).

When the objects are zones and the values correspond to aggregated data (values or counts), the detection of the local spatial association (hot spots and cold spots) between these aggregated data suggests that, locally, the scale of aggregation is not optimal and that a larger zone of aggregation would have allowed a better discrimination of the differences between zones. This subject will be addressed in the next chapter.

5.3.1.1. *Local Moran's index*

The local Moran's index [ANS 95] is a local decomposition of the global Moran's index. For each point P_i, the following can be written:

$$I_{Moran}(P_i) = \frac{1}{Si}\frac{(X_i - m)}{\sigma}\sum_{j,j\neq i} w_{ij}\,(\frac{Xj - m}{\sigma})$$

It is used for the characterization of the local autocorrelation and allows us differentiate statistically significant situations (by simulation, as in the general case, the simulation relates here only to the points involved in the local calculation). It is often associated with the calculation of a mean value weighted by the neighbors' values, compared to the value observed at point P_i.

5.3.1.2. *Getis–Ord index*

The formulation of the Getis–Ord index [GET 96] is similar to that of local Moran's index, but it yields a Z-score (Figure 5.8). For this, the local variance τ at point P_i is estimated and the index is:

$$G_{Getis-Ord}(P_i) = \sum_{j,j\neq i} \frac{w_{ij}}{\tau}(\frac{Xj - m}{\sigma})$$

Software	In practice
SavGIS	The calculation of local indices for spatial association is available with the menu: Calculer → Calcul géostatistique → LISA
ArcGIS	The calculation of local indices for spatial association is available in: ArcToolbox → Spatial Statistics Tools → Mapping Clusters
QGIS	The spatial autocorrelation tests are available with the menu: Processing Toolbox → SAGA → Geostatistics → Global Moran's I
R	Package spdep (localG)
GeoDa	The calculation of local indices for spatial association is available with the menu: Space → Local G Cluster Map

Table 5.10. *In practice*

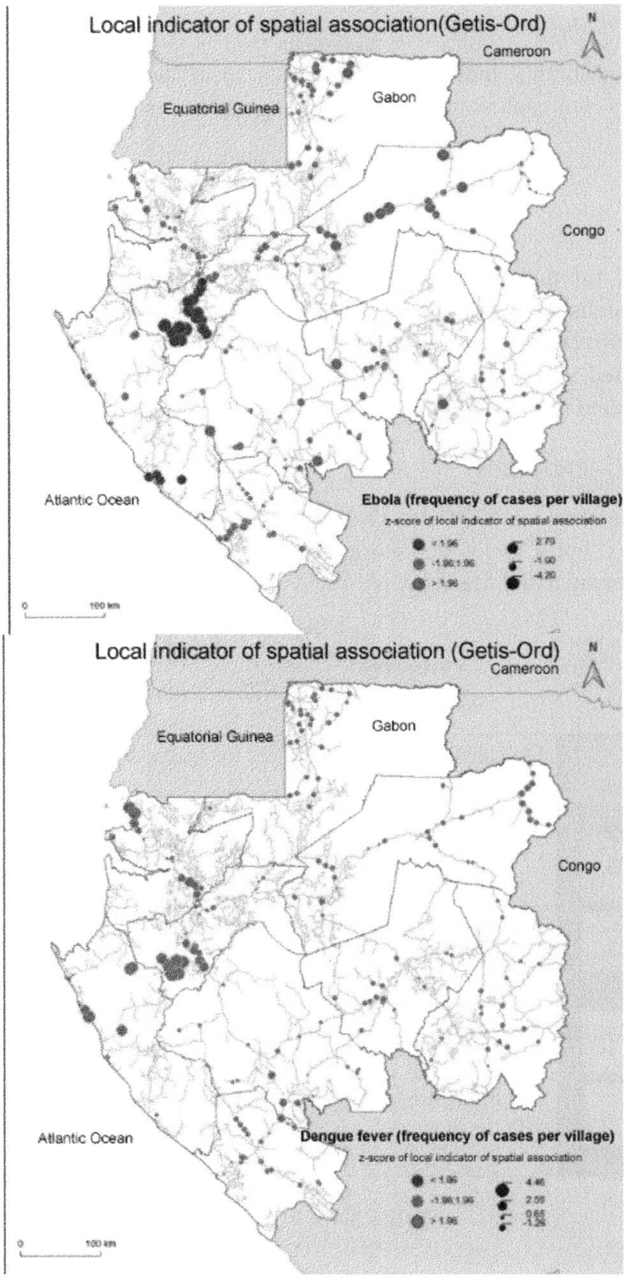

Figure 5.8. *Getis–Ord index on the serology values (Ebola, Dengue) from the data of a survey conducted in villages in Gabon. For a color version of this figure, see www.iste.co.uk/souris/epidemiology.zip*

5.3.2. *Spatial scan-based search for singularities*

Rather than characterizing the geographical objects themselves, the scan-based approach aims to search for points in space that are the place of a singularity with respect to the distribution of the set of objects or of a subset of objects among the set of objects (for example, infected farms among the set of farms). Several types of singularity can be expected: a cluster that is a significant grouping of events or values around a point, a centrality, which is an arrangement around the point that is particularly homogeneous in terms of distance, an attraction–repulsion, that is, the property of being significantly closer or further from events or values. Other tests can be built on the same principle, in order to study a particular spatial characteristic of a point in space with respect to a set of localized objects (events or values) [OPE 87].

A scan-based search involves systematic testing of all points in space with a given spatial gap, which amounts to testing each point of a grid.

Unlike methods characterizing objects themselves (villages, farms, towns, etc.), spatial scan-based methods are used for finding only the most statistically significant point(s) in a grid, in order to avoid the problem of multiple tests.

5.3.2.1. *The search for local clusters*

QUESTION.– Is there one or several clusters of events/objects/values? Where are the most significant cluster(s) located?

There is considerable interest in the search for clusters. It involves finding places that show a statistically significant clustering of events or a value (qualitative or quantitative – measure, quantity or proportion) among a set of objects [BES 91]. All of the methods rely more or less on the idea of counting the events or the values with or without a geographical window – a zone – whose size and location vary in space [OPE 87]. The difficulty resides, on the one hand, in defining the characteristics of this window, and on the other hand, in estimating whether a situation detected as abnormal is actually significantly different from a random situation.

A scan-based search for clusters is therefore essentially an approach by spatial aggregation of objects into zones. The objective is to find zones in which a number of events, a density, a prevalence or an incidence is significantly higher than the value expected in the case of a random spatial distribution (or following an *a priori* given distribution). The space is scanned by varying the location and size of these zones. The shape of the searched zones is fixed in advance: circular disk, elliptical disk, square, hexagon, etc.

The phrase "significantly different from a random situation" has a statistical meaning: it is a zone whose index (test statistics chosen depending on the problem to be solved) obtained by spatial aggregation of events or values from the global population allows us to reject the null hypothesis with a statistical test. The null hypothesis corresponds to a homogeneous distribution of events or values on the studied space (which means that the events or values are globally distributed according to a given uniform law of probability of parameter λ0) while the alternative hypothesis corresponds to the existence of a cluster in the distribution of events or values (the events or values inside the cluster are distributed with a different parameter λ1).

Since scanning creates many zones, testing all the zones raises a performance problem (the distribution of the test statistics under H_0 is obtained by Monte Carlo simulation) and a multiple tests problem. To solve these problems, the authors have proposed the selection of only some zones, using an increasing index depending on their probability of being a cluster (and on the significance of rejecting the null hypothesis) and the authors propose conducting the significance test only on the zones that maximize this index (these are known as *most likely clusters*). Therefore, the higher the index, the lower the probability of obtaining it by chance should be. Hence, if a zone does not allow the rejection of the null hypothesis, the zones with lower index values will not allow us to reject the null hypothesis either. This spares the effort of testing many zones and facing all the above-mentioned problems.

Following Kulldorf [KUL 95], several indices have been described, depending on the types of variables, the type of space (discrete or continuous) and an *a priori* distribution: Bernoulli models, Poisson models [KUL 95, KUL 97], multinomial models [JUN 10], ordinal models [JUN 06], normal models [KUL 09], exponential models [HUA 07]. These indices are built as a ratio between the likelihood of the observed situation (assuming different probabilities inside and outside the potential cluster and applying the principle of maximum likelihood) and the likelihood calculated from the model describing the distribution of events or values under null hypothesis (assuming the model is identical everywhere, and therefore the absence of a statistically significant cluster). The zones retained as potential clusters are those that maximize this index. The statistical significance of these zones is then calculated classically using Monte Carlo simulation to estimate the statistical distribution of the likelihood ratio, by randomly distributing the cases or depending on the *a priori* given process.

Index elaboration therefore requires expressing the likelihood as a function of the problem to be solved and the theoretical model chosen for describing it. As already seen, this model depends on the type of space (continuous or discrete), the type of process (homogeneous, non-homogeneous, dependent) and the type of variable (qualitative or numerical – quantity, count or measure):

Bernoulli model: the domain D is discrete (localized point patterns), and the variable V is Boolean (0/1) in each point. The Boolean variable may correspond to the choice of a modality if the initial variable is qualitative. This case is quite uncommon in practice, as it requires knowledge about the set of all susceptible individuals, their geographical location, as well as their state (yes/no). The dataset must therefore have comprehensive information on all the individuals, for example, case among non-cases (or control cases), infected villages among a set of villages and the health status of all trees in a set of trees.

This case can be extended to data already aggregated into geographical units, provided that not only the rate of positives is known, but also the number of cases and that of susceptible individuals in each unit, as all the counts must be known for the calculation of likelihood. The centroid of a geographical unit will be used as localization of cases and susceptible individuals contained by the unit. It is the most common situation in practice.

For this type of situation, the null hypothesis is modeled by a field of Bernoulli random variables of probabilities p identical in each point, while for the alternative hypothesis, a probability p' is assumed for the positives inside the potential cluster, which is higher than the probability q' outside the potential cluster. p' and q' are calculated from the observed data (applying the principle of maximum likelihood):

$p' = C_{in}/N_{in}$ (C_{in}: number of cases in the cluster; N_{in}: number of susceptible individuals in the cluster)

$q' = C_{out}/N_{out}$ (C_{out}: number of cases outside the cluster; N_{out}: number of susceptible individuals outside the cluster)

Null hypothesis: $p' = q'$

Alternative hypothesis: $p' > q'$

The likelihood of each situation is expressed by the Bernoulli variables corresponding to each event and is classically calculated by the product of their density functions:

$$K = p'^{C_{in}} (1 - p')^{N_{in} - C_{in}} q'^{C_{out}} (1 - q')^{N_{out} - C_{out}}$$

Bernoulli model can be extended to the case of a qualitative variable V with several modalities, which leads to what is known as the *multinomial model*.

Discrete Poisson model: the domain D is discrete (set of geographical units), the variable V corresponding here to a count of cases. The distribution of cases in the units under null hypothesis (H_0) is assumed to follow a Poisson distribution (and therefore a random distribution, with a number of cases proportional to susceptible individuals). The count of susceptible individuals must be known for each geographical unit in order to calculate the count of cases expected under null hypothesis. Unlike the Bernoulli model, where each individual has the same probability of being a case, the number of cases expected in a unit can be adjusted depending on another variable of the unit.

The likelihood is expressed here by $K = \left(\frac{C_{in}}{C_{exp}}\right)^{C_{in}} \left(\frac{C_{out}}{C_{in}+C_{out}-C_{exp}}\right)^{C_{out}}$, where C_{exp} represents the number of cases expected under the null hypothesis.

Space-time permutation or relative increase model: domain D is discrete, variable V is Boolean, similar to the Bernoulli model, but the susceptible population in each unit is not known. On the other hand, the date for each case is known. Assuming that the susceptible population does not change during the period, the emergence of geographical abnormalities in the increase in the number of cases can be studied; a cluster will be detected in a zone during a given period of time if the increase in cases during this period is significantly higher than the increase noted in the other zones during the same period.

Normal model: domain D is discrete, variable V is a numeric measure (body index, blood pressure, etc.).

Continuous Poisson model: domain D is continuous, and the events can emerge everywhere in the domain D. The cluster search concerns the location of events and not the number of events or the value of an attribute.

Restrictions can be applied to limit the eligible zones when searching for clusters: limitation of the number of events present in the cluster (with respect to the total number of events, for example, maximum 20%; in all the cases, the upper limit will be 50%), limitation of the number of significant clusters sought (the number is often two or three) and search for disjoint zones (no intersection between clusters).

The most common case is the use of disks as a cluster shape. Scanning then involves moving the center of the disk according to a regular grid (as a scanner) and varying the radius of the disk for each center. For each point of the grid, the value of the index is calculated for the set of disks centered on the point of the grid, whose radius increases to include events or objects up to the fixed limit (Figures 5.9 and 5.10). Only the disks that maximize the value of the index while meeting the chosen restrictions are retained.

Ellipses can also be used, especially if there is anisotropy of the spatial support (for example, villages along a river) or if anisotropy of the phenomenon can be assumed. Thus, the probability of detecting a cluster, if it exists, is higher; therefore, the power of the test is higher. Scanning involves varying the center of the ellipsis, the size and orientation of the major axis and for each major axis, the size of the minor axis. SaTScan™ software also proposes the addition of a criterion on the grouping of objects in the cluster, in order to consider the anisotropy while privileging the geographical concentration of objects in the ellipsis.

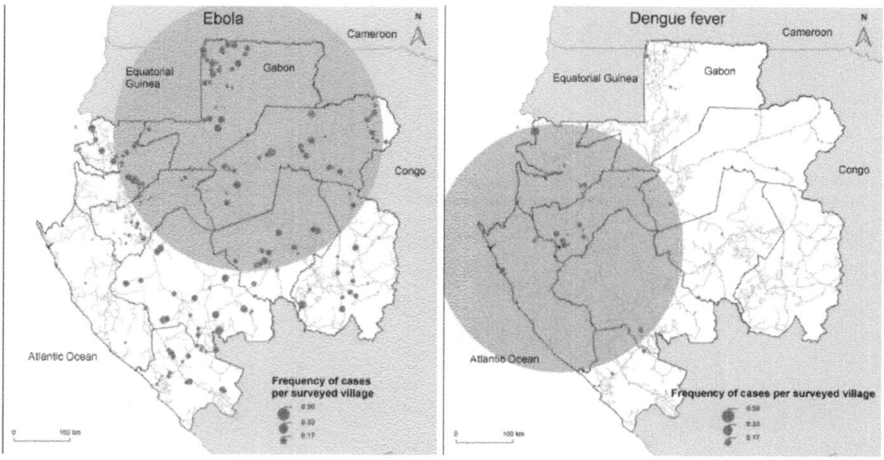

Figure 5.9. *Search for the most significant cluster: for Ebola seroprevalence (left), for Dengue seroprevalence (right) in the surveyed villages in Gabon. The location of the significant circular cluster of maximal value containing maximum 30% of the infected villages is represented here*

As will be seen in Chapter 7, the search for clusters can be extended to spatio-temporal data. Similar to the space–time permutation model, this search is conducted within a temporal window which varies in time.

SaTScan™ software, developed by M. Kulldorff [SAT 18], is the most complete software program for cluster detection.

SavGIS	The search for clusters by spatial scan is available with the menu: Calculate → Geostatistical calculation → Clusters
SaTScan™	SaTScan™ software specializes in the search for clusters by spatial scan

Table 5.11. *In practice*

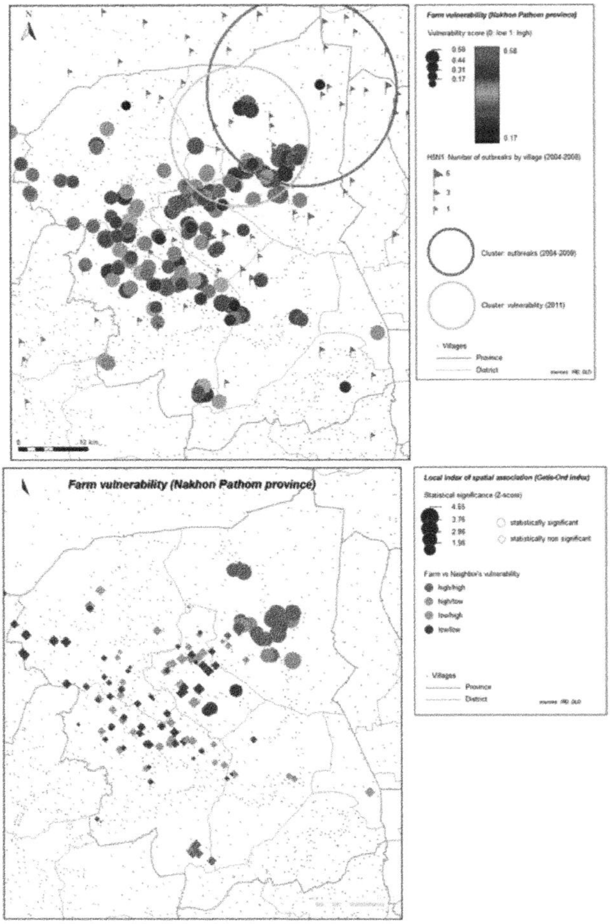

Figure 5.10. *Avian flu (H5N1) and the vulnerability of farms in Nakhon Pathom province, in Thailand: (top) comparison between a cluster of cases (2005) and a cluster of vulnerability (2011) shows that particular attention must be paid to a precise zone of the province. The Getis-Ord index of spatial association (bottom) confirms this analysis [SOU 14]. For a color version of this figure, see www.iste.co.uk/souris/ epidemiology.zip*

5.3.2.2. *Attraction–repulsion*

QUESTION.– Is there an interaction between points? What are the places that manifest attraction or repulsion for the studied phenomenon?

The objective is to determine each point P of space if the positive cases (for a Boolean variable in the point pattern) or the values (in the case of a numerical value) are significantly closer or further (stronger or weaker, respectively) with respect to this point. An index that reflects this property is built, such as in the Boolean case, for example, the median value of the distances from point P to positive points. Then, each point in space is scanned in order to highlight the point(s) with the highest and most significant values with respect to the random situation (points outside the confidence interval of the index, obtained by Monte Carlo simulation, by permutation of values among the points of the point pattern). For example, when the above-mentioned index is small, there is attraction, and when it is larger, there is repulsion (Figure 5.11).

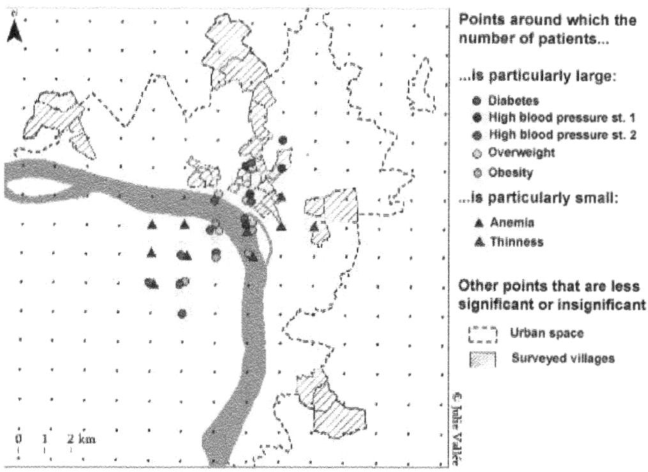

Figure 5.11. *Attraction–repulsion scan for several pathologies at Vientiane, Laos (according to J. Vallée).For a color version of this figure, see www.iste.co.uk/souris/epidemiology.zip*

SavGIS	Spatial scan-based attraction/repulsion is available with Calculate → Geostatistical calculation → Attraction/Repulsion

Table 5.12. *In practice*

5.3.3. *Analyses around a source point*

QUESTION.– Is a given place the center of a phenomenon? Is the distance to a place a risk factor for the studied phenomenon?

In the case of analyses around a source point, the analysis is limited to an *a priori* specified point. These analyses are common in the search for relationships between cause and effect around places assumed to be at risk (incinerator, nuclear power plant, chemical industry, antennas, etc.). In general, these studies focus on the identification of a cluster of events or excess risk around this source point.

The techniques used in the search for clusters are identical to those presented in previous sections, but they are limited to the source point considered. The objective is to highlight a zone (disk, ellipsis) around the source point that has a defect or an excess (in value or frequency, depending on the type of attribute or the studied event and the type of support, discrete or continuous).

In the case of a quantitative variable, the distance between each object and the source point can be calculated and a correlation between the value of the object and distance can be searched. Specific tests have also been developed for the trend analysis around a source point: Stone's test, Lawson–Waller's test, Bithell's test and Diggle's test [STO 88] [TAN 02] [PFE 08]. These tests aim to highlight an event decrease when the distance from a point increases (decreasing exposure).

The units are assumed to be classified from 1 to n with respect to their distance from the source point. The null hypothesis H_0 is: the risk is identical for all the units. The alternative hypothesis Ha is: the risk decreases as a function of the distance from the source point.

Stone's test uses the quantity:

$$T_{Stone} = max_{1 \leq j \leq n} \frac{\sum_{i=1}^{j} O_i}{\sum_{i=1}^{j} E_i}$$

Lawson–Waller's test uses the quantity:

$$T_{LW} = \sum_{i=1}^{n} g_i (O_i - E_i)$$

whose variance is given by:

$$Var(T_{LW}) = \sum_{i=1}^{n} g_i^2 E_i - O_i (\sum_{i=1}^{n} \frac{p_i g_i}{p})$$

where g_i represents the exposure to the source for an individual of the unit i, O_i is the number of cases observed in the unit i, E_i is the number of cases expected in the unit i, p_i is the population at risk of unit i and p is the total population.

The distribution of T under null hypothesis is estimated by Monte Carlo simulation.

Box 5.8. *Stone's test, Lawson–Waller's test*

In general, analyses around a source point go beyond a descriptive objective: they aim to establish cause-and-effect relationships [WAL 95, WAL 96]. It is therefore important to consider all the confounding factors that may influence the result of the analysis. Indeed, the fact that a case excess or a cluster around a point with potential risk has been highlighted does not amount to a cause-and-effect relationship at the level of individuals. The cluster and even the spatial trend can be due to spatial relationships between individuals and not to the relationship between the disease and the risk factor related to exposure (and therefore to the distance to a focus for the analysis around a source point). This is particularly true for contagious diseases: the cluster may be due to proximity-based transmission – mobile risk factor – and not to the presence of an immobile risk factor. More generally, it is difficult to conclude a cause-and-effect relationship with an immobile risk factor when the individuals are not spatially independent with respect to the studied disease (proximity-based transmission, family-related factors, socio-economic factors, influence of density, etc.). In order to conclude that there is a significant excess of cases due to an immobile factor present at one point (such as an incinerator or a nuclear power plant), the modeling of the null hypothesis needs to consider the clustering of cases due to mobile factors or to other immobile risk factors.

5.4. Example: emergence and diffusion of avian influenza

This example illustrates the use of tests for the statistical and spatial analysis of the emergence and diffusion of an epidemic, as well as the search for environmental risk factors. The objective is to determine the statistical and spatial characteristics of the epidemic in order to model it [SOU 10a, SOU 10b].

5.4.1. *Introduction*

The highly pathogenic avian influenza virus H5N1 remains a worldwide threat for human and animal health, while the mechanisms explaining its epizootic emergence and reemergence in poultry are to a wide extent still unknown.

Thailand experienced significant outbreaks of avian influenza due to H5N1 in 2000. The disease emergence processes and the relationships between emergence, diffusion and environment are studied here. A spatial approach is used for describing all the epizootic diseases from 2004 to 2008 and for the characterization of the emergence pattern, as two main scenarios are possible: multiple and independent introductions of the virus from the wildlife followed by moderate local propagation (due to agricultural practices) or, on the contrary, only one or very few emergences followed by a strong local propagation with rare spikes in long-distance diffusion, mainly due to the poultry farming and transportation practices.

The data used correspond to all avian influenza epizootic diseases signaled between July 2004 and February 2008 on the farms in Thailand. The basic study units are the infected farms, located at the level of their village (the precise location of the farms is not available). The farm cases were registered by the Farm Development Department (DLD, Ministry of Agriculture and Cooperatives in Thailand). These data correspond to the HPAI suspect cases detected by the passive and active clinical surveillance network (Figure 5.12). The data have been integrated into a spatial database managed by the SavGIS geographic information system.

Case-related data have been grouped in a file that lists the farms suspected of avian influenza in Thailand, from 2004 to 2008. Available information include: information on the farm, date of suspicion, code of the village, poultry type, total number of animals, total number of dead animals and total number of animals to be destroyed. A laboratory result allows us to identify farms that have tested positive for the virus.

	Number	date	day	month	year	Province Code	District Code	Sub-District Code	Village Code	Poultry type	Total	Sick	Dead	To Kill	Date of result	Lab Result
2	224	3/7/2547	3	7	2004	14	1408	140815	14081507	Egg Chicken	11327		3000	8327	7/7/2547	+
3	738	6/7/2547	6	7	2004	65	6501	650106	65010604	Domestic Chicken	8					-
4	738	6/7/2547	6	7	2004	65	6501	650106	65010604	Farm Duck	78		2			-
5	226	7/7/2547	7	7	2004	13	1301	130112	13011201	Egg Chicken	850		120	730	7/7/2547	+
6	228	9/7/2547	9	7	2004	64	6407	640708	64070805	Egg Chicken	600		23	577	9/7/2547	+
7	229	9/7/2547	9	7	2004	53	5308	530805	53080506	Egg Chicken	1570		484	1106	9/7/2547	+
8	232	9/7/2547	9	7	2004	64	6407	640712	64071211	Domestic Chicken	100		20	80	9/7/2547	+
9	233	9/7/2547	9	7	2004	64	6405	640509	64050902	Domestic Chicken	225		35	190	9/7/2547	+
10	235	9/7/2547	9	7	2004	49	4904	490404	49040407	Domestic Chicken	2083			2083	12/7/2547	-
11	237	12/7/2547	12	7	2004	41	4104	410404	41040412	Domestic Duck	300		110		14/7/2547	-
12	238	12/7/2547	12	7	2004	41	4109	410906	41090601	Domestic Chicken	100		25		14/7/2547	-
13	240	12/7/2547	12	7	2004	60	6002	600209	60020906	Farm Chicken	8000			2500	12/7/2547	+
14	274	12/7/2547	12	7	2004	13	1305	130507	13050711	Domestic Chicken	100		30	70	12/7/2547	+
15	274	12/7/2547	12	7	2004	13	1305	130507	13050711	Farm Chicken	5430			4810		+
16	281	12/7/2547	12	7	2004	13	1304	130403	13040309	Farm Chicken	1200			1200	12/7/2547	+
17	286	12/7/2547	12	7	2004	15	1507	150704	15070404	Egg Duck	18400			18400	12/7/2547	+
18	384	12/7/2547	12	7	2004	13	1305	130506	13050605	Farm Chicken	10000	0		10000	12/7/2547	+
19	391	12/7/2547	12	7	2004	72	7205	720503	72050301	Egg Duck	4701	0	4701		12/7/2547	+
20	391	12/7/2547	12	7	2004	72	7205	720503	72050302	Egg Duck	4701	0	4701		12/7/2547	+
21	396	13/7/2547	13	7	2004	64	6407	640708	64070804	Domestic Chicken	1087			1000	13/7/2547	+
22	404	13/7/2547	13	7	2004	64	6407	640708	64070806	Domestic Chicken	4955			1643	13/7/2547	+
23	408	13/7/2547	13	7	2004	57	5701	570120	57012007	Domestic Chicken	472		55	417	13/7/2547	+
24	413	13/7/2547	13	7	2004	15	1506	150604	15060402	Egg Duck	3500			3500	16/7/2547	+
25	227	14/7/2547	14	7	2004	67	6706	670604	67060411	Domestic Chicken	35			35	23/7/2547	-
26	231	14/7/2547	14	7	2004	40	4007	400710	40071008	Farm Chicken	600		185	415	19/7/2547	-
27	417	14/7/2547	14	7	2004	62	6204	620412	62041204	Domestic Chicken	2000			65	14/7/2547	+
28	421	14/7/2547	14	7	2004	13	1303	130306	13030602	Farm Chicken	4500		245	4255	14/7/2547	+
29	424	14/7/2547	14	7	2004	16	1601	160110	16011005	Farm Chicken	14000		4179	9871	14/7/2547	+
30	424	14/7/2547	14	7	2004	16	1601	160110	16011005	Egg Chicken	22		22			+
31	424	14/7/2547	14	7	2004	16	1601	160110	16011005	Domestic Chicken	40		25	15		+
32	424	14/7/2547	14	7	2004	16	1601	160110	16011005	Fighting Cock	78		50	28		+
33	426	14/7/2547	14	7	2004	19	1901	190107	19010707	Farm Chicken	8000		1	6866	14/7/2547	+

Bird Flu

Figure 5.12. *The list of farms suspected between 2004 and 2008 comprises more than 10,000 farms, among which 1,807 tested positive for the H5N1 virus*

Only the farms with poultry deaths due to biologically confirmed avian influenza are of interest here. The list will therefore be narrowed down using the result of the laboratory confirmation test. This test yields a positive result when the presence of H5N1 virus has been highlighted. A negative result does not necessarily mean that the virus is absent, only that it could not be detected: false negatives may have been eliminated, but the test provides the possibility of excluding false positives. Following this operation, 1,807 cases were retained.

5.4.2. *Mapping*

5.4.2.1. *Case mapping*

The farms' location uses the coordinates of the village (centroid) to which they belong. It is possible to visualize this new layer of information. The farms belonging to the same village have the same location and overlap on this map. Similarly, if a farm has been contaminated several times, it is listed several times and the points overlap. Therefore, this mapping is not adequate. The number of cases is aggregated by village, sub-district, district or province (Figure 5.13).

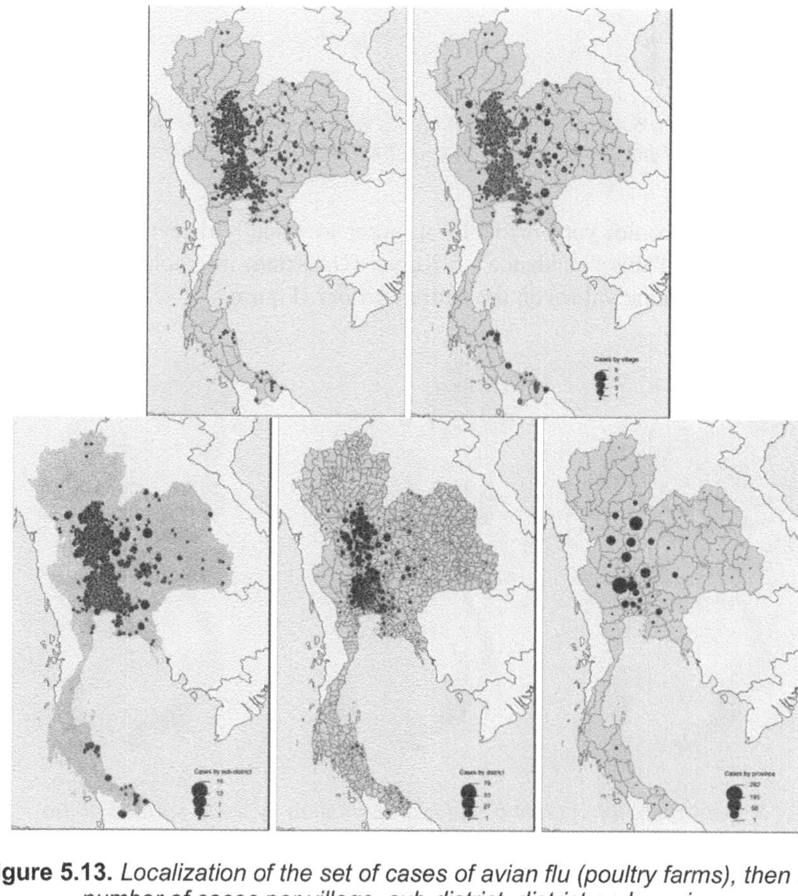

Figure 5.13. *Localization of the set of cases of avian flu (poultry farms), then the number of cases per village, sub-district, district and province*

The number of poultry farms by sub-district is available. It is thus possible to calculate and map the proportion of cases of avian flu with respect to the number of farms by sub-district, district and province (Figure 5.14).

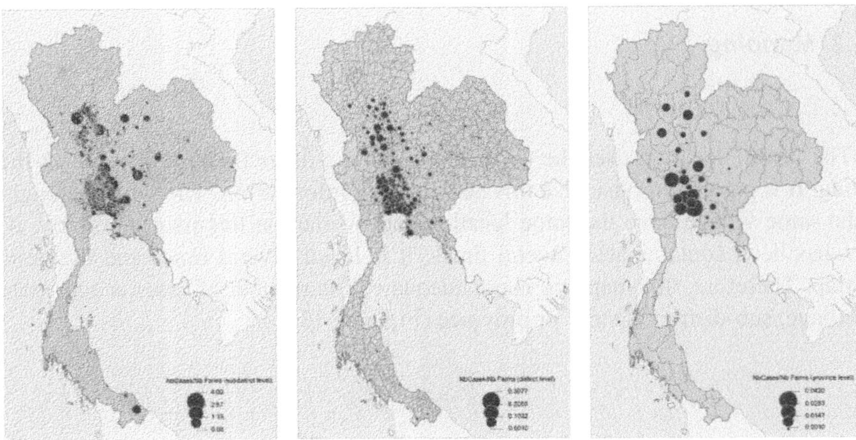

Figure 5.14. *Proportion of cases with respect to the number of poultry farms, by sub-district, district and province*

These maps are not very readable. In order to visually detect the spatial trends and the clusters of this "incidence", a Kernel (Gaussian) interpolation can be made in order to smooth the values on the entire territory (Figure 5.15).

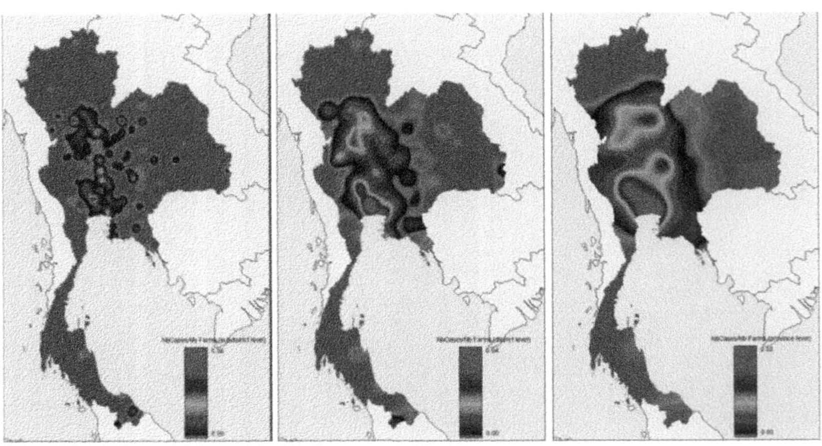

Figure 5.15. *Interpolation and representation by trend surface of the proportion of cases with respect to the number of poultry farms, by sub-district, district and province. For a color version of this figure, see www.iste.co.uk/souris/epidemiology.zip*

5.4.2.2. *Thematic mapping*

Below is a representation of the percentage of farms infected depending on the poultry type, divided into three categories: chicken, duck and other (goose, quail, turkey, etc.). A representation with sector-based diagrams is used, where size is a function of the number of farms infected in the province (Figure 5.16).

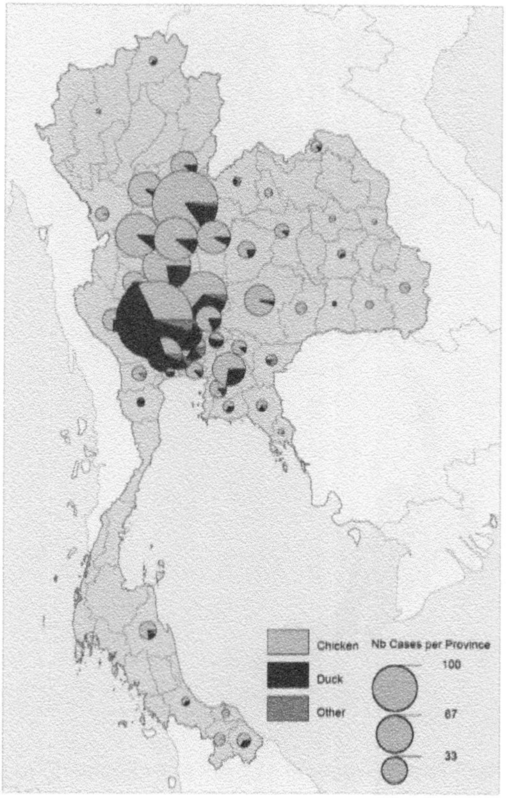

Figure 5.16. *Representation by province of the percentage of infected farms depending on the type of poultry raised at the farm*

5.4.3. *Analysis of the spatial distribution of cases*

5.4.3.1. *Centrographic analysis*

For an in-depth analysis and in order to compare the spatial distribution of infected places and uninfected places, a two-dimensional centrographic analysis is made by calculating the centers of gravity and the standard deviational ellipses for the infected and uninfected villages (Figure 5.17).

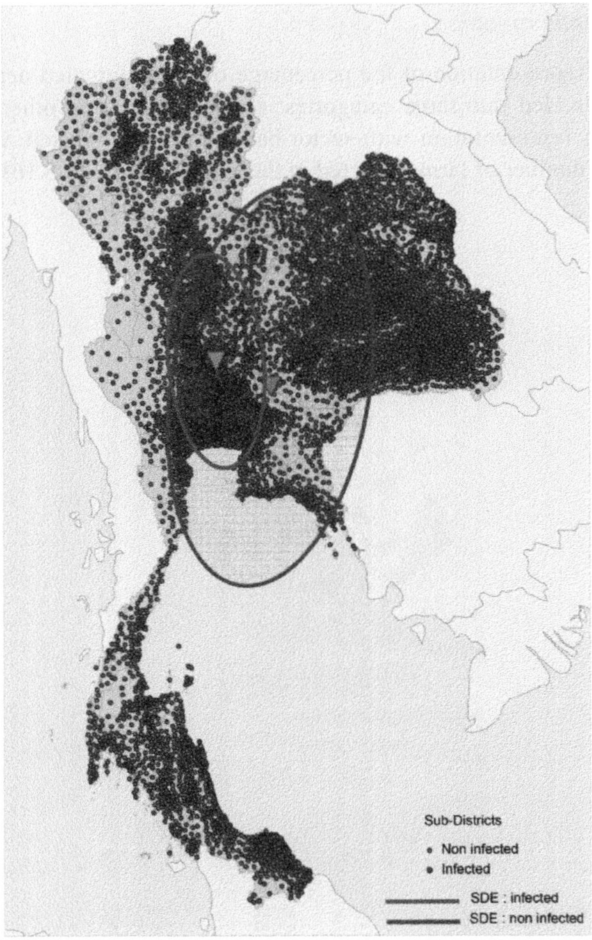

Figure 5.17. *Standard deviational ellipses corresponding to the location of infected villages (red) and uninfected villages (green). For a color version of this figure, see www.iste.co.uk/souris/epidemiology.zip*

5.4.3.2. *Analysis of the global spatial distribution*

Even though it is visually apparent that the ellipses differ in terms of location, pattern and size, statistical tests must be conducted to formally prove that the spatial differences between the two sets are not due to random fluctuations. First, the randomness of the mean geographical location of the infected villages with respect to uninfected villages can be tested. The real values must be compared to the values obtained by simulation, which allows us to evaluate the random variability of the index used.

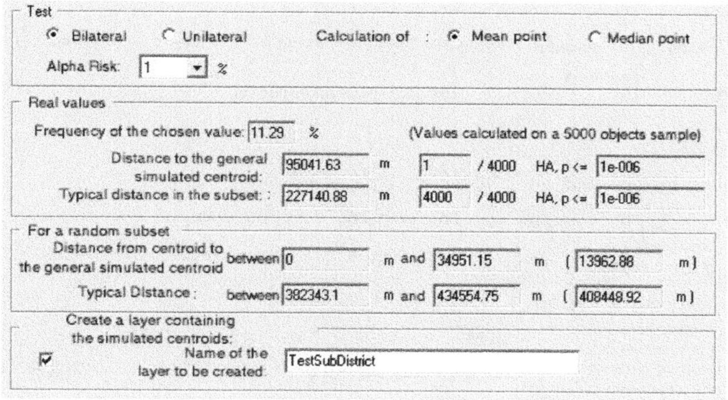

Figure 5.18. *Statistical results for global location analysis. Please note that this figure has been translated into English from French*

The test result shows the strong statistical significance in the rejection of the null hypothesis. The visual result shows that the location of the centroid of the observed cases is very distant from the cloud of the simulated centroids (Figure 5.19).

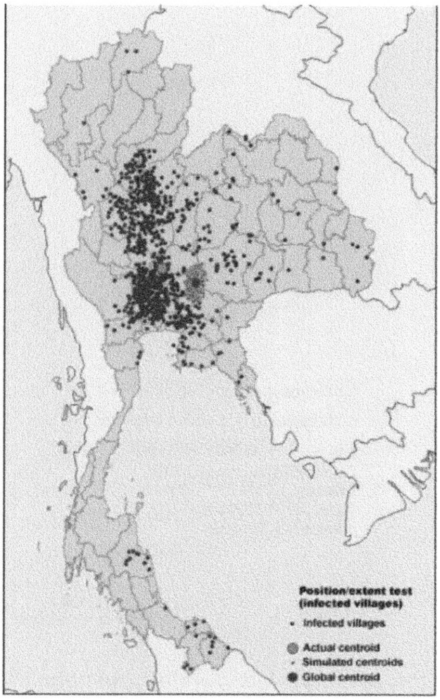

Figure 5.19. *Location of the centroid of infected villages (orange), cloud of centroids of simulated situations (sky blue) and centroids of simulated centroids (dark blue). For a color version of this figure, see www.iste.co.uk/souris/epidemiology.zip*

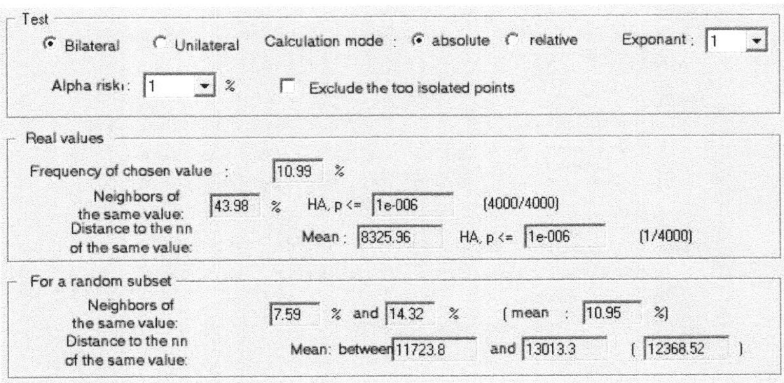

Figure 5.20. *Result of the spatial distribution of infected sub-districts in the set of sub-districts. The observed value (44% of infected sub-districts have their nearest neighbor also infected, compared to 11% for a random distribution) shows a strong clustering of infected sub-districts. Please note that this figure has been translated into English from French*

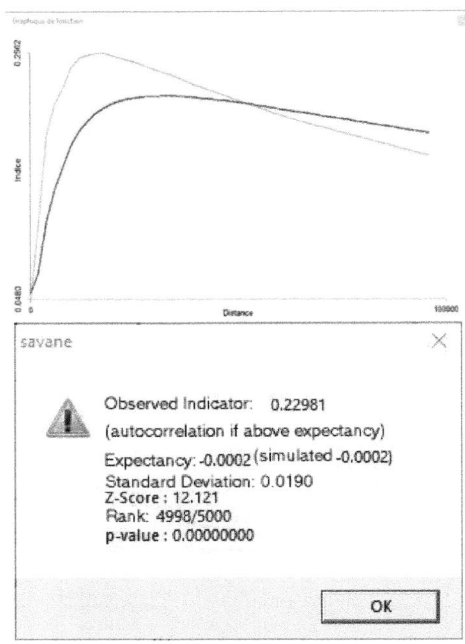

Figure 5.21. *Value of the autocorrelation index (number of cases by sub-district), depending on the maximum distance for spatial weights (left: green curve and yellow curve with Souris-Demoraes correction), and significance of the index for a maximum distance of 20 km (right). The minimum distance between two sub-districts is 10 km on average. Please note that this figure has been translated into English from French*

The following test allows the evaluation of the overall clustering of infected villages. The objective is to verify that the distribution of infected villages is not random and that this distribution is more clustered than if it was due to chance, which is to be expected for a disease that is *a priori* transmitted by contagion and therefore probably by proximity from one farm to another. The first test uses the percentage of infected villages whose nearest neighbor is also infected, as well as the mean distance between an infected village and its infected nearest neighbor, comparing the values observed to the values obtained by simulation when the infected villages are randomly distributed in the set of villages. The second test uses an index of spatial autocorrelation of the number of infection cases by sub-district, and a strong autocorrelation can also be noted, which means that the close sub-districts have more similar values than distant sub-districts (Figure 5.20). The analysis of the autocorrelation value as a function of the maximal distance in the calculation of weight provides information on the intensity of autocorrelation, which is maximal for a research radius of 20 km (Figure 5.21).

5.4.3.3. *Analysis of the local spatial distribution*

Our objective is to analyze the local differences observed between sub-districts in order to highlight the places or the values that are significantly different from those of their neighbors and to measure the intensity of local clusters. The operation allows us also to classify places depending on their value compared to that of their neighbors (high–high, low–high, high–low, low–low).

A local index of spatial association (Getis-Ord) is used here over the number of cases of avian influenza by sub-district. For the spatial weight, a linear decrease from 1 to 0 is used up to a distance of 30 km, and 0 beyond that. The calculated index, which is a Z-score indicating the significance of the difference between a point and its neighbors, can be visualized by a symbol or in the form of trend surface, after interpolation (here Kernel Gaussian). Moran's scatter plot indicates that most of the significant sub-districts are within "+, +" quadrant (values above the mean with values of neighbors above the mean). The plot is centered on the mean values. It can also be noted that the clusters of values that are significantly lower than expected, and consequently in the "–, – " quadrant, are close to the mean and not very significant, unlike the clusters of significantly higher values, which are displayed all over the "+, +" quadrant (Figure 5.22).

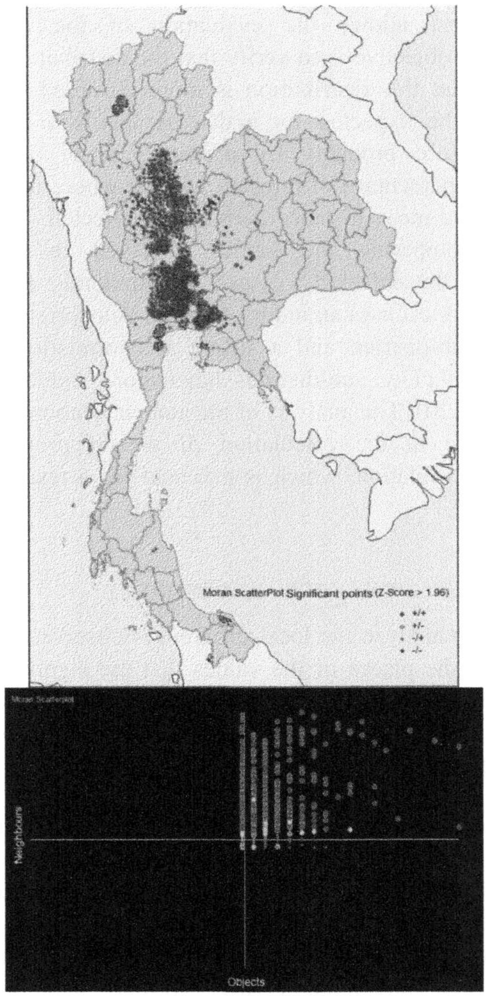

Figure 5.22. *Map of significant sub-districts (Z-score > 1.96) depending on their class (++, +−, −+, −−); map of Z-score of Getis-Ord index and Moran scatter plot. For a color version of this figure, see www.iste.co.uk/souris/epidemiology.zip*

When the spatial distribution of cases shows a clustered distribution, it is natural to search for the places, if they exist, that concentrate most cases. The objective is to highlight the existence and the location of the most statistically significant clusters of cases. This search for spatial cluster involves "spatial statistical scan" procedures. The analysis by sub-district allows us to highlight three clusters of sub-districts presenting cases of avian influenza (when the number of sub-districts per cluster is limited to 30% of sub-districts having at least one case) (Figure 5.23).

A priori, the presence of places of case clusters does not mean that the phenomenon is non-stationary, but that the case generation process is contagious. The characteristics of this process can be evaluated from the characteristics of the observed clusters. To highlight a non-stationary process that could explain the presence of a cluster in one place or in a region, an environmental element (in the broad sense) that characterizes these places or this region must be highlighted (type of farm, infrastructures, land cover, etc.).

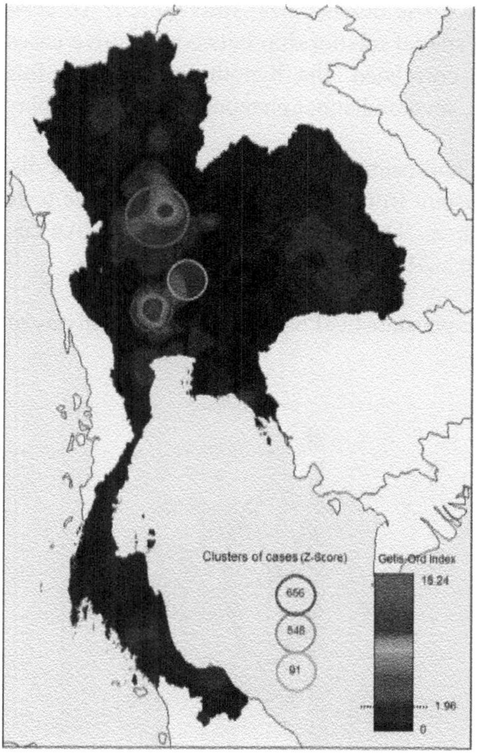

Figure 5.23. *Circular clusters that are the most significant for the number of cases per districts. The calculation requires clusters to contain maximum 20% of the cases. For a color version of this figure, see www.iste.co.uk/souris/epidemiology.zip*

The statistical analysis of infected sub-districts that are in the clusters compared to those that are outside the clusters shows no significant difference for different environmental variables (land cover, infrastructures, population, etc.). This result already suggests that the clusters at the level of sub-districts are due to proximity-based diffusion rather than to an environmental or socio-economic risk factor.

5.4.3.4. *Comparison of the distribution of positives and negatives*

In the context of the control of avian influenza, many cases of abnormal mortality have been reported by the system of veterinary surveillance. The samples taken are analyzed for the presence of the H5N1 virus. Among these samples, many are negative. How should the relationships between positives and negatives be analyzed? Are most negatives false negatives (effectively positive for avian influenza, but with a negative test result), or are they caused by another disease (Gumboro virus, Newcastle disease, heat stroke, etc.)? To answer this question, we can study the global spatial relationship between positive cases and negative cases. If the two sets are correlated, the hypothesis of many false negatives can be advanced. Otherwise, several distinct phenomena may be at play.

The cartographic and centrographic analysis indicates a distribution of negative tested cases that is distant from the distribution of positive tested cases, but close to the set of declared cases. Therefore, it seems that negative cases are not related to positive cases (Figure 5.24).

To confirm this, a geostatistical test based on the analysis of bivariate colocation will be used.

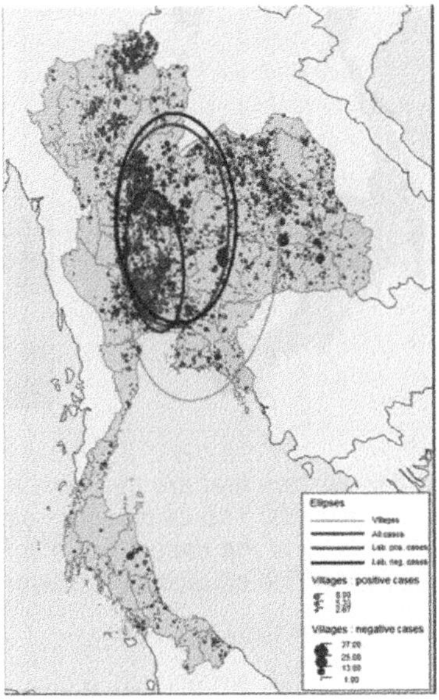

Figure 5.24. *Reported cases that tested positive (red) and tested negative (blue). The ellipses represent the synthesis of the location and extent of the set of villages (orange), the set of reported cases (green), cases that tested negative (blue) and cases that tested positive (red). For a color version of this figure, see www.iste.co.uk/souris/epidemiology.zip*

Both the villages with cases that tested positive and the villages with cases that tested negative show very clustered spatial distributions ($p < 0.000001$). The colocalization index LS*, conducted on the villages that reported at least one case, shows a negative spatial correlation, regardless of the distance *d*0 used for the calculation of the spatial weight. The colocalization index based on the nearest neighbors also indicates this result, which is reinforced when a clustering condition is imposed to the simulated sets. Therefore, the two sets do not correspond to similar phenomena; consequently, two distinct phenomena exist, the reported cases that tested positive having a spatial distribution uncorrelated to that of the reported cases that tested negative (Figure 5.25).

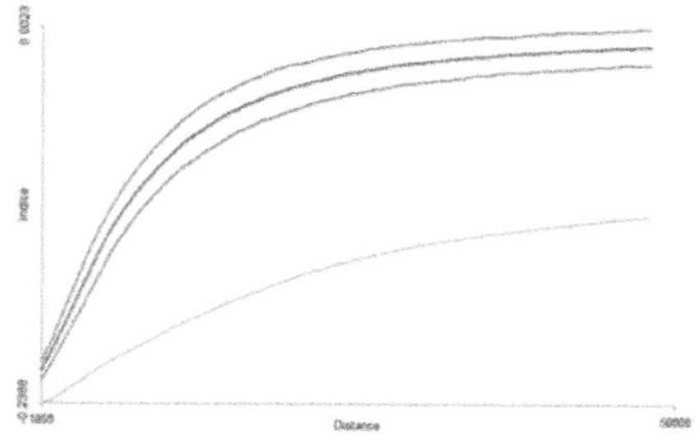

Figure 5.25. *Index LS* is significantly below the value obtained by simulation (green, with a risk of error of 5% in gray), regardless of the maximal distance of influence (bandwidth): the spatial distributions of the two sets are significantly different. For a color version of this figure, see www.iste.co.uk/souris/epidemiology.zip*

5.4.4. *Spatio-temporal analyses*

5.4.4.1. *Mapping*

The temporal distribution of cases can be easily visualized on a histogram of the number of cases depending on time (Figure 5.26).

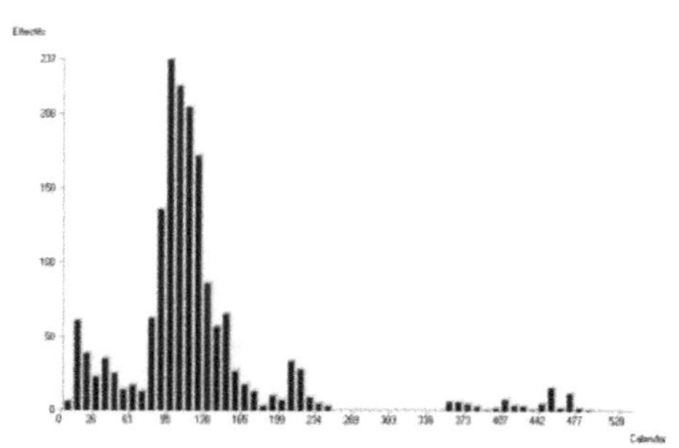

Figure 5.26. *Number of cases that tested positive in time for the two first outbreaks of epizootic disease in Thailand*

A spatio-temporal animation is made by mapping the occurrence of cases as a function of time and then assembling the images to create a video (Figure 5.27).

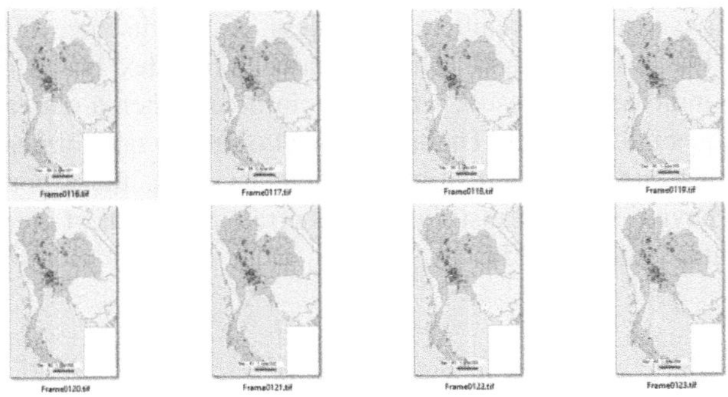

Figure 5.27. *Several images of the time sequence. The set of images will be assembled to create a video showing the occurrence and the accumulation of cases in time*

It is also possible to visualize a mobile mean of cases, depending on the date. The visual analysis of the distribution gives a suggestion on the movement of the center of the epidemic. Referring to the avian influenza outbreak, it can be noted that the mean moves first towards north, then returns to the initial location, and there are no periods when the mobile mean remains immobile (Figure 5.28).

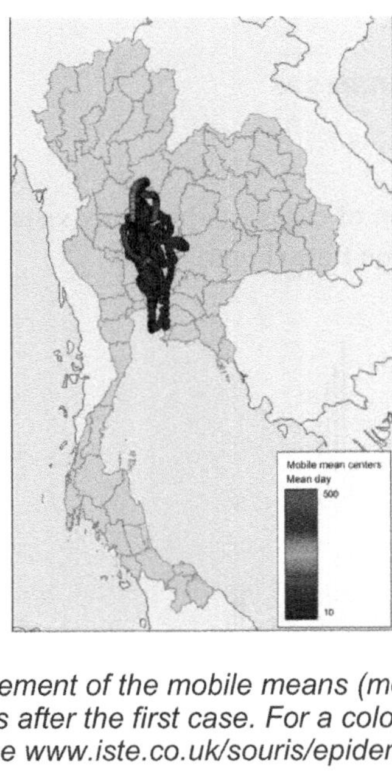

Figure 5.28. *Movement of the mobile means (means over 10 days) up to 500 days after the first case. For a color version of this figure, see www.iste.co.uk/souris/epidemiology.zip*

5.4.4.2. *Spatio-temporal statistical analyses*

Statistical analysis on the date on which the cases are reported shows a global spatial autocorrelation of time: cases are closer in space when they are closer in time, which is a logical expectation in the case of a contagious disease. This result is also confirmed by the Knox test (Chapter 7), which shows a relationship between case proximity in time and case proximity in space as soon as the distance between cases exceeds 5 km, regardless of the difference in time between two successive infections (Figure 5.29).

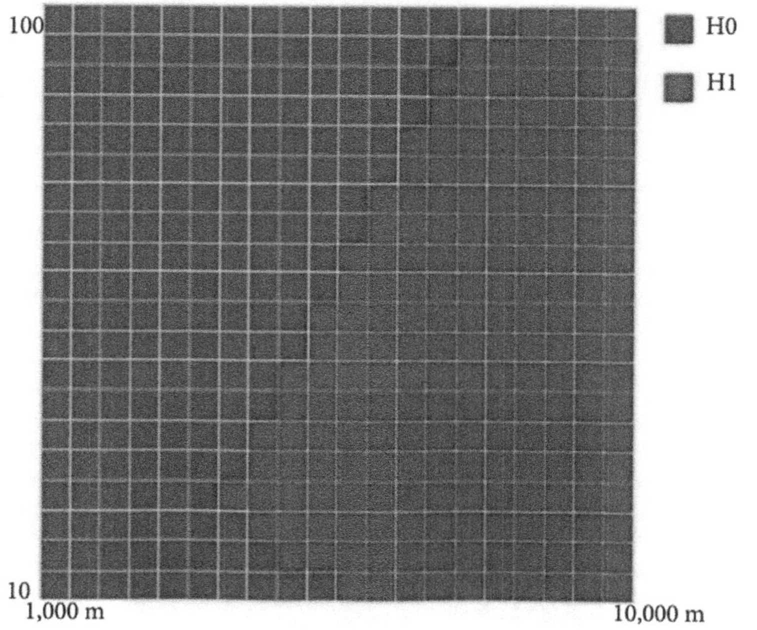

Figure 5.29. *Matrix of Knox tests. x-axis: the distance between infected places; y-axis: the number of days between two successive infections*

The local indices of spatial association highlight the spatial disparities in the distribution of an index of spatial autocorrelation in time (Figure 5.30). These analyses should be treated with caution, since the various temporal episodes may mix, and their separation involves considering the various epidemic outbreaks that can be detected, either visually on a histogram or automatically by searching for natural thresholds in the distribution. Later on, it will be shown how this problem can be directly addressed by reconstructing the pathway of the epidemic.

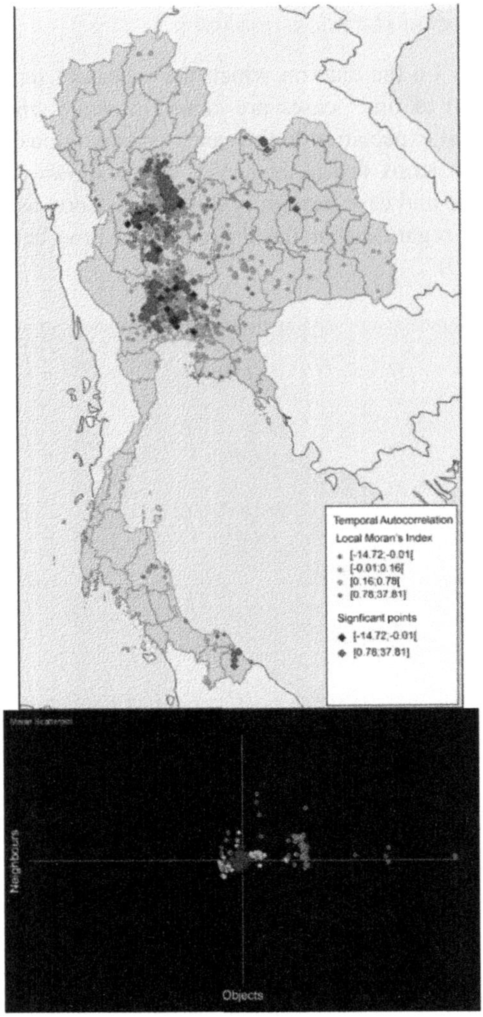

Figure 5.30. *Local Moran's index for the date of infection and the corresponding Moran scatter plot. The points in red on the map correspond to villages where infections are most correlated in space and time. Statistically significant values are represented by a rhomboid (type I risk at 5%). For a color version of this figure, see www.iste.co.uk/souris/epidemiology.zip*

Finally, an analysis of the presence and location of a cluster during time allows the visualization of the movement of the main site of the epidemic. Using a 20-day time window that is moved every 10 days, a significant main cluster is obtained, which moves from the initial site of the epidemic to the north, comes back close to its initial location and then heads towards north (Figure 5.31).

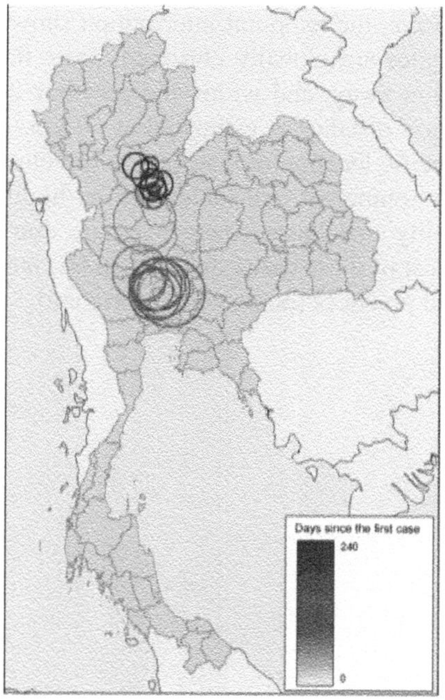

Figure 5.31. *Movement of the main significant cluster of cases over time, up to 250 days after the start of the infection, corresponding to the first epizootic wave (20-day window with a displacement of 10 days between each cluster calculation). For a color version of this figure, see www.iste.co.uk/souris/epidemiology.zip*

5.4.4.3. *Emergences*

The objective here is to separate emergence processes from diffusion processes, which allows, on the one hand, the detection of environmental risk factors for the occurrence of the disease and, on the other hand, the characterization of the disease propagation mechanism in order to choose between two scenarios: multiple introductions of the virus followed by moderate local diffusion or, on the contrary, very rare emergences followed by strong local diffusion and several rare jumps at long distance. A spatio-temporal filter will be used for the separation of emergence from proximity diffusion: a case is called emergent if it has no precedent within a given lapse of time and a given radius. The first scenario corresponds to a repeated introduction by vectors or wild reservoirs, such as migrating birds, followed by weak proximity-based diffusion between farms. The second scenario rather corresponds to very rare introduction and reemergence (from the wildlife) and to a very dynamic proximity-based diffusion due to exchanges between farms and sporadic diffusion at longer distance due to poultry transportation (in principle, very controlled since the start of the epidemic).

For each epizootic wave, the temporal interruption shown by the cases does not exceed one week. The sites are globally clustered: more than 90% of cases had a precedent less than 10 km away, and within the previous 21 days. If the maximal proximity-based contagion distance is estimated at 60 km, the cases considered as emerging display no specific location or distribution (no clustering, no spatial trend) (Figure 5.32). If no environmental risk factor can be highlighted in the points of emergence, the probability that these emergences are due to the wildlife can be considered to be low: most of these cases should then be considered as long-distance jumps due to commercial practices (poultry transportation).

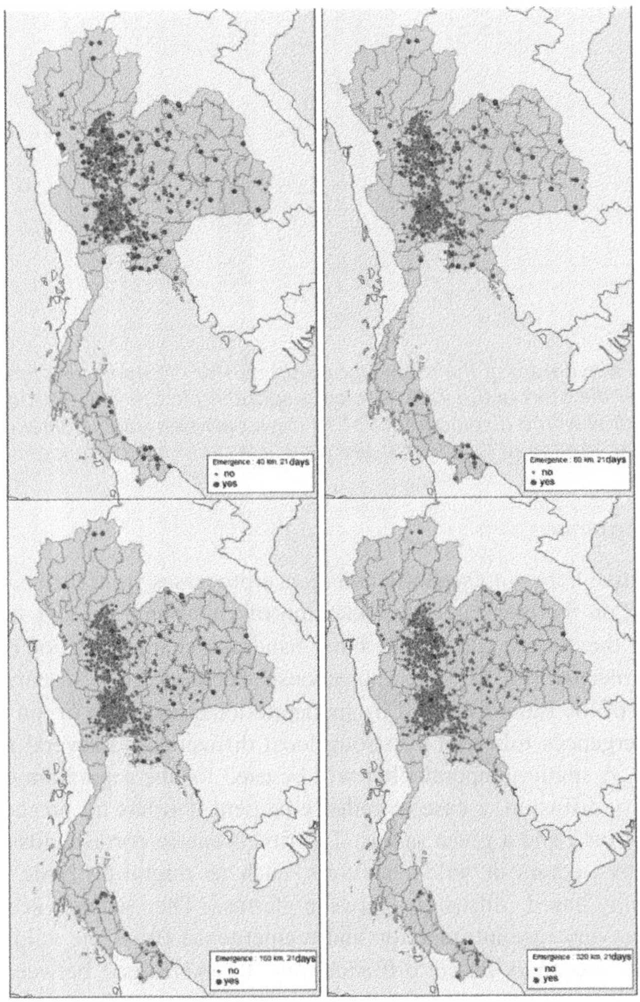

Figure 5.32. *Selection of emergence points, depending on various criteria of distance and time*

5.4.4.4. *Pathway reconstruction*

Once the points of emergence have been determined, a model of the epidemic pathway can be established, and each case can be assigned the point of emergence to which it corresponds. Pathway reconstruction is done by backward chaining depending on time and distance, until all the points are linked to a point of emergence (Figure 5.33).

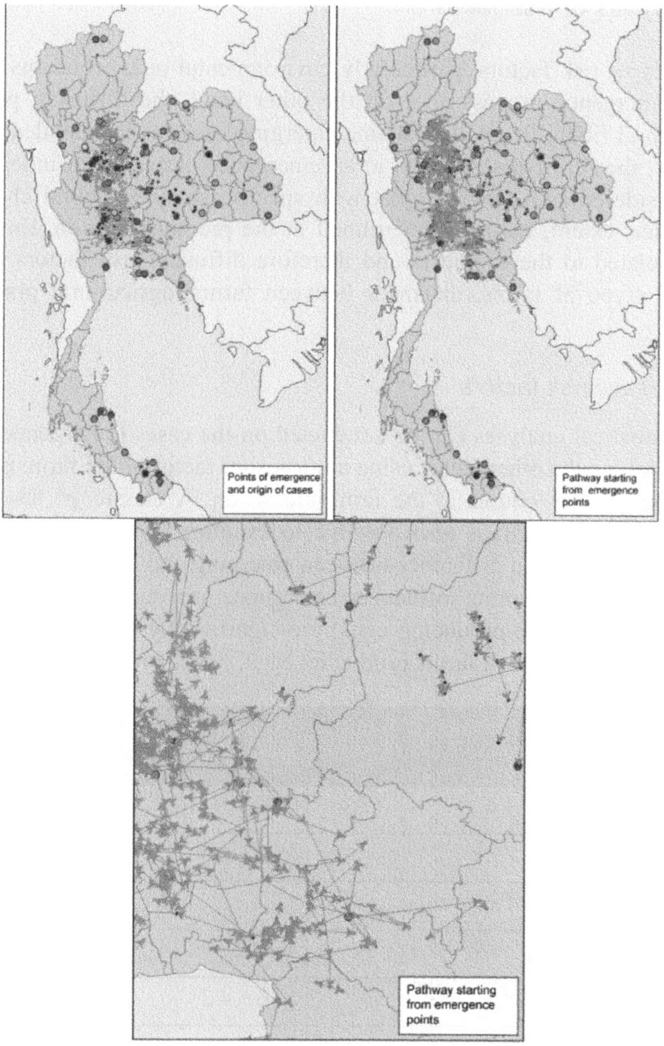

Figure 5.33. *Modeling of the origin of cases depending on the points of emergence and reconstruction of the pathway of the epidemic. For a color version of this figure, see www.iste.co.uk/souris/epidemiology.zip*

All these results allow us to estimate the diffusion parameters (notably density and spatial distribution of cases around a point of emergence, speed of movement, extent and so on. All these parameters will then be used for modeling the diffusion process around a point of emergence, the environmental risk factors of which will be determined in the following section.

5.4.5. *Analyses of risk factors*

The study of risk factors, particularly environmental ones, concerns, on the one hand, the emergence process and, on the other hand, the diffusion process. The spatio-temporal filter allows us to define emerging cases as independent events. On the contrary, the set of cases linked to an emergence case are not independent and will be considered as a single object with specific spatio-temporal characteristics (speed, center, extent, etc.), as determined in the previous section. Environmental factors correlated to these objects and therefore diffusion risk factors can also be investigated (type of farms, distances between farms, agricultural practices, land cover, etc.).

5.4.5.1. *Intrinsic risk factors*

Many statistical analyses can be conducted on the cases (emergence on the one hand, diffusion on the other hand) using an exposure factor of the farm: poultry type, farming type, poultry counts in the farm and so on. A classic χ^2 test or Fisher's exact test will be used. It is also possible to examine, for all these intrinsic risk factors, the existence of a difference between emerging and non-emerging cases. For example, the type of farming influences emergence: among infected farms, poultry farms (particularly those producing eggs) are significantly more emergent than the other farms, unlike the duck farms producing eggs, which are less emergent:

Poultry type	"Emergent" farms (%)	"Non-emergent" farms (%)	p-value
Bantam cock	1.6	0.5 ± 0.9	0.15 (H_0)
Domestic chicken	59.2	55.3 ± 6.2	0.24 (H_0)
Domestic duck	1.6	1.1 ± 1.3	0.36 (H_0)
Egg chicken	10.9	4.4 ± 2.5	0.005 (Ha)
Egg duck	4.7	15.6 ± 4.2	0.007 (Ha)
Farm chicken	9.4	4.9 ± 2.6	0.049 (Ha)
Farm duck	3.2	7.6 ± 3	0.081 (H_0)

Farm francolin	3.2	2.2 ± 1.8	0.30 (H_0)
Fighting cock	3.2	2.7 ± 2	0.42 (H_0)
Free-range duck	1.6	1.1 ± 1.5	0.47 (H_0)
Goose	0	0.8 ± 1	0.23 (H_0)
Mandarin duck	1.6	2.5 ± 1.9	0.30 (H_0)
Turkey	0	0.2 ± 0.6	0.34 (H_0)

Table 5.13. *Statistical results for emergent and non-emergent farms*

5.4.5.2. *Environmental risk factors*

It is impossible to conduct an on-site analysis of the risk of infection due to environmental factors, as the full list of farms is not available. On the contrary, an agricultural census at sub-district level is available. It is therefore possible to analyze the environmental risk factors on these geographical units. A description of land cover throughout Thailand is available. It was established in 2000 by the Thai Royal Forest Department from high-definition satellite images, and therefore, it has the accuracy required by this calculation at the level of sub-districts. The environmental factors will be evaluated by aggregation or geographical juncture in sub-districts (sums, percentages, surface percentages).

On the contrary, it will be possible to test the differences between "emergent" cases and "non-emergent" cases at farm geographical level, since these analyses are limited to infected farms. Environmental factors will be evaluated by defining a buffer zone around each infected farm (Figure 5.34).

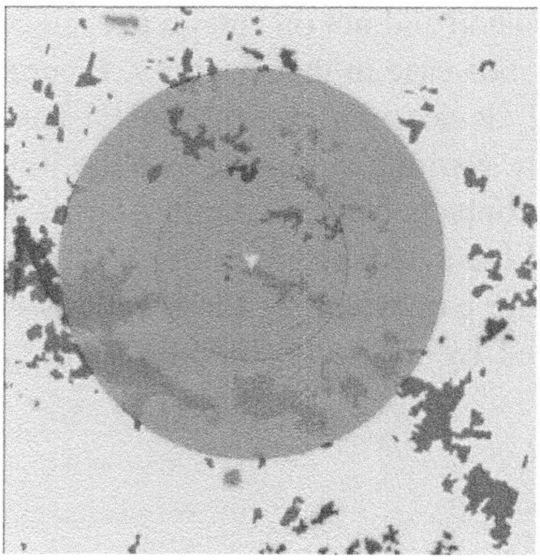

Figure 5.34. *Buffer zone around a farm (200 m, 400 m) for the calculation of immobile environmental risk factors (water surface percentage)*

The geographical units for which the risk of infection is zero are eliminated, as there is no susceptibility (no farms). An "emergence" attribute (yes/no) is assigned to sub-districts depending on the presence of emergences in the sub-district: the sub-districts that had at least one emergent case are called "emergent sub-districts". The number of non-emergent cases by sub-districts and by point of emergence allows us to also measure diffusion intensity by sub-district. The statistical analysis of exposure uses classic tests of comparison between the two groups (exposure to risk factors between "emergent" sub-districts and "non-emergent" sub-districts, with possible consideration of a confounding factor) as well as correlation tests (intensity of diffusion and percentage of the surface filled by the studied risk factor).

For example, one of the main hypotheses usually advanced to explain the emergence of avian influenza is that of pathogen introduction by migratory birds. To test this hypothesis, the water surface percentage per sub-district was calculated, as migratory birds' rest is strongly linked to the presence of water bodies. These analyses show no relationship between the presence of water bodies in the sub-district and the presence of emergence, which reinforces the probability of long-distance diffusion due to commercial practices rather than repeated introductions of the virus by wildlife.

The analysis of differences between emergent and non-emergent cases shows that the characteristics of the emergence differ from the characteristics of diffusion: for example, emergence is significantly more probable in the farms raising chickens than in the farms raising ducks, but no difference is noted for diffusion. On the contrary, no environmental differences are noted between "emergent" and "non-emergent" cases.

In conclusion, this study shows that the effort for preventing avian influenza should focus on local diffusion and not on emergence, since most cases seem to have a relationship with a previous case in their vicinity. Emergence remains a rare event that is difficult to control, as no environmental factor has been highlighted. On the contrary, an efficient early warning system allows the limitation of the diffusion to the short range (banning contacts between poultry farms and trying to eliminate the virus around the site) and the avoidance of jumps to longer distances (restricting all poultry transportation). The parameters of these actions can be evaluated from the diffusion model elaborated by the reconstruction of the pathway of the epizootic disease.

SUMMARY.–

– Direct analyses of the spatial distribution of objects or values can be global or local.

– Most of the methods for spatial analysis use in principle statistical tests and simulation in order to estimate the distribution of the test statistic.

– Global spatial analyses concern location, extent, orientation, centrality and spatial dependence of values.

– Global indices of spatial autocorrelation allow us to explain the spatial dependence of values in the entire studied domain.

– Bivariate spatial analyses allow us to compare global spatial distributions of two phenomena.

– In order to study the local variations of spatial autocorrelation, a local index of spatial association (LISA) is used.

– Spatial scan methods are used for highlighting distinct places with respect to the spatial distribution of the phenomenon: clusters, attraction–repulsion, centrality, etc.

– The studies around a source point aim to highlight excess risk around the point.

6

Spatial Analysis of Risk

6.1. Introduction

This chapter presents spatial analysis methods that rely on the statistical analysis of variables previously obtained by spatial calculations, such as aggregation of events into geographical units or the spatial belonging of an event. Spatial calculations can be classified into two groups: spatial aggregates, which generate new variables in the geographical aggregation units, and spatial belongings, which generate new variables for the individuals using their spatial belonging to geographical units.

Aggregation and belonging are frequently used to generate new variables after which direct methods of spatial analysis can be used, as described in the previous chapter. It is also very common to use aggregation and spatial belonging methods in geographical analyses, such as accessibility or influence analyses or the analysis of care supply and use.

6.2. Aggregation-based spatial analyses

The methods presented here concern the statistical and spatial analysis of the result of an aggregation into geographical units. These *indirect* methods do not directly involve the spatial relationships between individuals, but only belonging to the same space. This characteristic distinguishes them from the *direct* methods presented in the previous chapter. They are nevertheless very widely used, particularly because they allow the assessment and analysis of probabilities and risks.

The analysis concerns geographical units, not individuals. The initial individuals are aggregated into geographical units (already existing or generated for the occasion) using their geographical belonging to these units. The analysis then studies these aggregation units with classic statistical methods (for the distribution of values between aggregation units) and spatial methods (for the analysis of the spatial distribution of the aggregation units' values). The analyzed data relate either to the mean values of individuals (mean per aggregation unit of a numeric value of the aggregated individuals), or counts (sum per unit of aggregation of a number of individuals, count of a modality of a qualitative variable or sum of a numeric value representing counts), or frequencies, sometimes variances.

As a first stage, classic methods are used to analyze aggregation results, without considering the spatial relationships between aggregation units. In the second stage, direct spatial methods are used to analyze the spatial distribution of aggregation units, using the position of their centroids or their topological relationships (proximity, adjacency, etc.).

The questions that can be addressed using descriptive statistical analysis, which will be detailed in the following sections, are mainly the following:

– Are there any geographical units whose value is "abnormally" high or low? What are the units that have these "abnormal" values?

– What is the extent and statistical distribution of aggregated values? Is this extent compatible with a given distribution (such as a random distribution)? Does the distribution of values between the various geographical units present an abnormal variance (overdispersion or underdispersion)?

– Is the geographical division employed adapted to the problem raised? What is the aggregation scale allowing the maximization of differences between units?

– Is there a risk factor correlated with the variable to be explained? Is it possible to find a statistical model that explains the values observed from variables of the geographical units? Does this model need to consider the values of neighboring units, and how? Does this model need to consider spatial relationships between individuals, and how?

The questions that spatial analysis must address are partly identical to those examined in the previous chapter:

– Is the spatial distribution of values compatible with a given distribution, such as a random distribution or a distribution with specific characteristics (stemming from a specific process, such as a contagion process)?

– Does the spatial distribution present specific spatial characteristics (autocorrelation, gradient, shape, direction, centrality, clusters, etc.)?

– Is it suggesting the influence of an environmental risk factor?

– If they exist, do the significantly high (respectively low) values or the residue of a statistical modeling present a specific spatial distribution (such as autocorrelation, cluster, direction, etc.)?

Before a detailed description of the methods used to address these questions, several considerations on the spatial aggregation process and its consequences for statistical and spatial analysis are worth presenting.

6.2.1. *Spatial aggregation operation*

As already highlighted, health events are rarely available at the agent level (individuals, vectors, reservoirs, etc.) for either technical or ethical reasons. In particular, the accurate location of an event is rarely available, while a large part of the phenomena occur and are explained at the level of agents and relationships between agents (genetic factors, susceptibility, exposure, contacts, contagions, etc.). Obtaining exhaustive spatial information at this level is very often impossible. A possible option is to decide to ignore the actual location and to replace the agents by new objects, grouping the (mobile) agents according to their belonging to immobile geographical units, whose location is well defined, as zones resulting from administrative or geographical division. Another possible option is to aggregate the events according to their belonging to geographical units that already have other data. This is applicable especially when attempting to establish statistical correlations between the events and these other variables (so-called "environmental" correlations, mesological analyses) and to assess risks. A final possibility is to aggregate the agents into geographical units devoid of geographical significance, of identical shape and size, as square or hexagonal unit cells or disks around points. In addition, GIS operations make it possible to calculate environmental data for these units (see Appendix, available online at: www.iste.co.uk/souris/epidemiology.zip).

Although not directly involving spatial relationships between individuals, the aggregation operation implicitly brings in the proximity between individuals, as they are aggregated according to their belonging to spatial units that are most often interconnected. If the intensity of these spatial relationships between individuals is correlated with population density, the latter variable could be used as a risk factor in the statistical analysis at the level of geographical units.

The second stage (the spatial analysis itself) essentially uses the techniques described in the previous chapter, with the customary restrictions, since these are aggregation units, mean values and not the initial individuals.

When the initial data are already aggregated into spatial units, as it often happens in epidemiology, the receiving units are obviously imposed by the data themselves. This aggregation is *imposed* and not *chosen*. If required, a new aggregation in a division generalizing the initial division can be considered.

In general, the geographical units of aggregation are zones, but they can also be lines or points; it is always possible to define a distance $d0$ in order to aggregate objects located at a distance below *d*0. This distance is generally null for the zones, aggregation using the geometrical belonging to the zone. For points and lines, this distance is generally above 0. All of these operations of spatial aggregation either into grids, zones, lines or points are available and easy to implement in geographic information systems (see Appendix, available online at: www.iste.co.uk/souris/epidemiology.zip).

Aggregation is interesting for many reasons: it replaces several objects (individuals or events) with only one, replaces values with means or counts, and thus reduces random variability of the phenomenon at the level of individuals. Aggregation units are generally immobile, which significantly simplifies spatio-temporal analyses. Aggregation makes it possible to add epidemiological attributes to units that already have values (for example, population, surface, type of soil, etc.) and to use these values to create new attributes, such as rates or densities. Therefore, it allows the assessment of probabilities in the geographical units, and hence the spatialization of a situation by disaggregating a global estimation. It allows the comparison between receiving units in order to describe and analyze the determining factors of possible statistical or spatial differences, or to search for correlations with other attributes. Varying the size of the aggregation units, it allows the analysis of the phenomenon at several scales and the search for optimal scales for the identification of spatial structures.

Aggregation of events or values into (immobile) geographical units transforms individual data, events or agent (known as "emitter") variables into variables of these geographical units, called "aggregation units", "grouping units", "receiving units" or simply "receivers". These new variables are counts, mean values, frequencies and variances.

There are two types of aggregation:

– aggregations that use existing geographical units, such as administrative divisions. In the emitting object, belonging is often given by the name or code of the receiving unit, but it can also be calculated if the precise location expressed in coordinates is available for emitters and receivers;

– aggregations that use the creation of geographical units, such as a grid of identical cells or disks around points. It is possible to choose the shape or size of the cells, and to analyze several aggregations by varying the size of cells or disks. This second type of aggregation is not possible unless the location of the events to be aggregated is available in coordinates, with an accuracy that exceeds the grid size.

In "classic" epidemiology, the evaluation of the statistical distribution of a variable in a population uses the values of a set of individuals, considered as so many random variables (independent and identically distributed in the population). When spatialization is considered, this set of random variables becomes a set of regionalized random variables; however, the population as a whole (the process) cannot be spatially characterized based on these random variables unless strong stationarity hypotheses are emitted. Aggregation allows an intermediate approach.

Box 6.1. *Spatial aggregation*

The comparison between units of aggregation raises many problems related to accuracy, scale, significance, context and confounding factors. These problems are referred to as MAUP (Modifiable Areal Unit Problem) [OPE 79]. The aggregation process has many consequences. When individuals are replaced by groups of individuals and individual values are replaced by mean values or densities, individual information is lost, being replaced by statistical information on a group. In particular, it leads to the loss of all information on spatial relationships between individuals, the direct use of these spatial relationships between individuals in the comparative analysis of groups being impossible.

When statistical information on the group is designed as an assessment of a probability for this group, the quality (possible bias) and the variance of this assessment must be studied. Using these probability assessments and comparing them may be challenging for several reasons: when assessment accuracy is not the same in each group (accuracy depends on denominator, see below), when there is a significant internal variation in the aggregation units (improper characterization of geographical units with respect to the studied process), when temporal relationships are neglected (consideration of events that belong to different processes) and when aggregation leads to loss of information that is essential for understanding the phenomenon (for example, low probability and strong impact events or relationships between individuals, or the influence of a context of another spatial level, such as density). The quality of the characterization of an individual phenomenon provided by aggregation must be evaluated and discussed. For example, the dispersion of individual values inside receiving units must be analyzed when studying the representativity of a mean.

Another problem occurs when, instead of aggregating point events, data that have already been grouped in a finer division are aggregated in a division; when the two divisions are not hierarchical, this operation involves solving the problem of non-hierarchical divisions. Methods for virtual distribution of individuals from one division into another have also been implemented in GIS.

When individual data are aggregated into a geographical unit, it is often possible to know a population assigned to this unit, and therefore to calculate rates (incidence rate, prevalence rate, density rate, risk rate, etc.). These rates are often used implicitly for the assessment of probabilities and risks. Caution is nevertheless recommended when using ratios. On the one hand, the accuracy of a ratio depends on a denominator, which differs from one aggregation unit to another, which is often ignored subsequently. On the other hand, a ratio at the group level is the expression of a phenomenon that may differ from the phenomenon studied at the level of individuals. When each aggregation unit is characterized by a ratio, a linear relationship (proportionality) is implicitly assumed between aggregated data and denominator, as if the probability of the phenomenon at the level of individuals was automatically transmitted at the group level. This hypothesis may be false if the individuals are not independent with respect to the studied phenomenon. For example, the probability of an individual changing its sanitary status may depend on the status of other individuals (this is particularly true for infectious diseases: the probability of an individual being sick increases if a neighbor is sick). Many phenomena are related to population density, and calculating a ratio leads to losing this information. If a known individual risk factor does not have the same statistical distribution in various units of aggregation, differences between units depend on these differences in distribution; these differences should be eliminated if we want to eliminate the influence of this risk factor. This is the object of standardization, already described in Chapter 4, and which will be detailed later in this chapter.

Let N be a set of individuals (susceptible population), out of which n are positive (Boolean variable observed on the individuals). Each individual i has a theoretical probability Pi of being positive. The simplest case is $Pi = constant$, for any i (equiprobability of individuals), but sometimes we have a model allowing the assessment of individual probability: $P_i = f(X1, \dots, Xn, \varepsilon)$, where Xj is an individual or environmental variable (such as age, gender, profession, place of residence, quality of air, etc.). Sometimes all we know is that the probability depends on variable Xi, the mathematical expression of this dependence being unknown (the hypothesis of a piecewise linear function is made in order to adjust the consequently expected counts – it is the object of standardization).

The phenomenon is assumed to originate in only one process. The model allowing the assessment of individual probability is therefore calculated from the complete dataset.

Let G_k be a set of spatial units. Individuals i are aggregated into units G_k by spatial belonging. N_k denotes the number of individuals of G_k, and n_k denotes the number of positive individuals of G_k.

If the probability of each individual being positive is considered independent of the other individuals, the theoretical count of positive individuals in each group G_k follows a binomial law whose mean P'_k and variance V_k can be easily calculated from individual probabilities P_i and group count (in case of equiprobability P of individuals P_i for every i, then $P'_k = P, V_k = N_k P'$). The condition of independence between individuals with respect to the health phenomenon is often not satisfied, particularly for infectious diseases. The binomial distribution should then be replaced with a distribution that takes the non-independence of individuals into account.

Box 6.2. *The process of spatial aggregation of a Boolean variable: count data*

The search for correlations at the level of the receiving unit also presents problems. The aggregation process results in the loss of the direct relationship between individual and exposure to risk factor, and the correlation result concerns the receiving unit, and not the individuals that compose it. The aggregation process consequently bans any conclusion involving this direct relationship, which would lead to what is referred to as "ecological fallacy" (see Chapter 2). It is therefore forbidden to infer at the level of individuals a result that has been obtained from a group, unless group variance is null.

Finally, the characteristics of domain D, spatial support of the phenomenon (the individuals), disappear to the benefit of the characteristics of domain D', spatial support of aggregation units. These characteristics can be very different. Very often, there is a passage from a continuous domain (for example, possible movement of individuals) to a discrete, point domain (such as villages), or from a discrete, point domain, to a continuous domain (zones representing the entire space).

Given all of these reasons, it is essential to clearly specify the objects being studied, as well as the variables used, and to express the choices and hypotheses associated with them.

SavGIS	Spatial aggregation calculation is available with the menu: Quest → Geoaggregation

Table 6.1. *In practice*

6.2.2. *Statistical analysis*

Statistical analysis involves the analysis of the statistical distribution of values of the receiving units (extent, dispersion, distribution, extreme values, etc.) and the statistical modeling of the observed values depending on unit variables (considered as risk factors).

The general approach, the elements of which will be detailed in the remainder of this chapter, is the following:

– values are calculated in each receiving unit: counts, means, prevalence and incidence rates (if the population at risk is known for each object), relative risk and odds ratio (when an exposure factor can be quantified in each object, quantification being often obtained by spatial aggregation with another layer of information);

– to eliminate known individual risk factors, aggregated data are standardized by means of data on the population structure in the aggregation units;

– the distribution of aggregated data is studied: extent, dispersion, highlighting of units that present "abnormal" values, inter- and intra-group variance;

– the correlations between values and environmental data are studied;

– the objective is to find a general model that explains an observed variable (mean, count or ratio) based on explanatory attributes.

SavGIS	Many statistical calculations are available in Stat menu, among which the menu: Stat → Epidemiology

Table 6.2. *In practice*

6.2.2.1. *Aggregation and standardization*

An important pitfall of aggregation concerns the known individual factors that are assumed to influence the phenomenon at the individual level and are found in the comparison of groups. This factor acts as a confounding factor in the analysis of group comparison: when individual data are aggregated into groups and these groups are compared by calculating risk measures (prevalence, incidence rate), it is often necessary to eliminate the differences in distribution between the groups of a known individual factor, in order to eliminate (or minimize) the influence of this factor on group differences. It is the object of *standardization* [BOU 93].

The main objective of health status mapping in a geographical division based on individual data is often to visualize spatial differences between these units. Known risk factor standardization is therefore essential if the objective of the map is to highlight the spatial differences independently of this factor.

The typical example is age standardization (for many diseases, infectious or not, susceptibility or vulnerability is age-related), but any other variable can be used for standardization. Age will be used in the following, but it can be replaced by gender, ethnicity, level of education, etc.

If a pathology known to be age-related is studied in two populations, the differences in age distribution between the two populations directly influence the difference in rates between the two populations, and the observation of rate differences reflects the differences related to disease and the differences related to population age structure. The relationship between age and disease is obviously not known directly; otherwise in the calculation of rates during aggregation, it would be sufficient to replace equiprobability of individuals with the probability of being sick depending on age, making a weighted calculation. When an aggregation has been made, it is unfortunately impossible to go back to the individual characteristics of the individuals. Data standardization requires knowledge on the distribution of the standardization variable in the geographical units.

Direct standardization involves approximating this age function by a linear function on each class of age. It is sufficient to know the coefficients in each interval in order to adjust the number of positive individuals depending on the structure of a typical population. To adjust the count of a geographical unit, the coefficients of each class of age are calculated from the data observed in this unit. Standardization must be conducted on the set of geographical units (Figure 6.1).

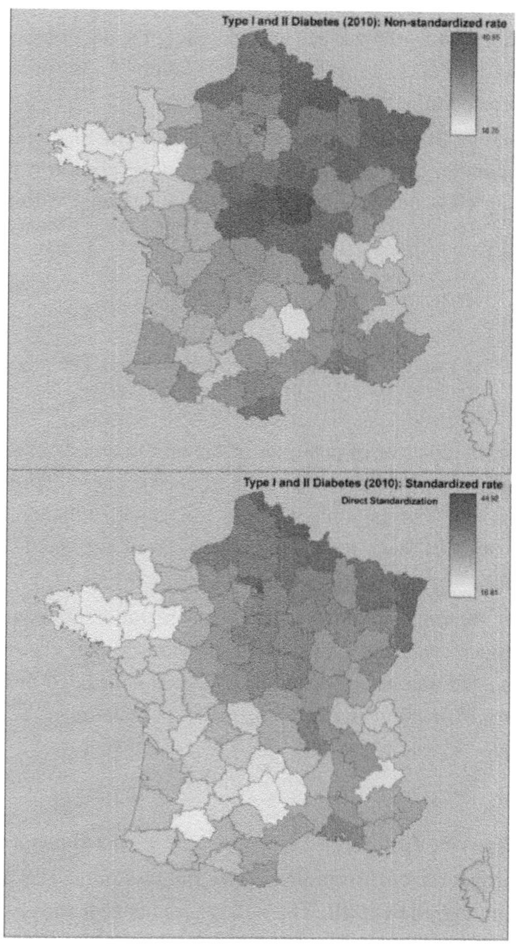

Figure 6.1. *Diabetes in France, 2010. Mapping of non-standardized rate in the upper image and of standardized rate in the lower image (direct of standardization age). The south-west/north-east trend is much clearer with standardized rates. For a color version of this figure, see www.iste.co.uk/souris/epidemiology.zip*

As it generally follows a process of disaggregation of the global dataset into spatial units, then into classes of the confounding factor (age, for example), the counts of each class in each unit are often small, and the direct standardization methods cannot be used, as they lack in accuracy. The indirect method should be used, even though it means to evaluate the rates per class of age in the entire dataset (standardization is then referred to as *internal*). Paying attention to the comparisons made is important, because it is difficult to compare standardized ratios when population structures are very different (see the following section).

> Direct standardization involves evaluating for each aggregation unit the coefficients per class of the standardization variable (such as age), using the data observed in the unit, then applying these coefficients to a reference population R in order to calculate the standardized rate corresponding to the unit:
>
> Let O_i be the number of cases of the unit in class i, N_i be the population of the unit in class i, N be the population in the unit, R_i be the count of class i in the reference typical population R. p is the class number. The count O' standardized with respect to R for the unit is:
>
> $$O' = \sum_{i=1}^{p} \alpha_i R_i$$
>
> where $\alpha_i = {O_i}/{N_i}$ is the rate observed for the unit in class i. The standardized rate of the unit is ${O'}/{R}$.
>
> The number of individuals or of positives of the unit in a class of age is sometimes too small and does not allow the accurate estimation of coefficient α_i. Direct standardization is not applicable in this case. For example, in order to evaluate with an accuracy of $0.5*10^{-3}$ an incidence whose order of magnitude is 10^{-3}, the class of age should contain a minimum of 15,000 individuals.
>
> When the counts are too small, the indirect standardization method can be used. It uses β_i rates that have been accurately calculated previously per class of age over large populations. An expected number of positives E is then calculated in each unit depending on these coefficients and on the structure of the actual population of the unit:
>
> $$E = \sum_{i=1}^{p} \beta_i N_i.$$
>
> The ratio ${O}/{E}$ between the observed count and the expected count is denoted as SMR, which stands for Standardized Morbidity or Mortality Ratio. The adjusted rate is the observed rate ${O}/{N}$ multiplied by SMR. The differences between the observed count and the expected count can also be studied directly.

Box 6.3. *Direct standardization and indirect standardization*

Standardization does not solve the problems of accuracy of the estimations of coefficients, when the counts of populations used for these estimations are very different or when the counts per class are very different.

EXAMPLE.– Let us consider one young person and 100 aged persons. Let us calculate the mean value of the 100 aged persons in order to standardize this population, and have the same weight for the value of the young person and for the value of the aged persons. The mean value of the 100 aged persons is much less variable than the value of the young person; therefore, the standardized value depends too much on the value of the young person.

SavGIS	Standardization is available in the menu: Calculate → Statistical calculation → Standardization

Table 6.3. *In practice*

Standardization can be replaced by a global statistical modeling approach, which allows the consideration of several factors as well as the interactions between these factors (age and gender, for example). Working on the residues after modeling makes it possible to address group comparison problems.

6.2.2.2. *SMR (Standardized Mortality Ratio)*

Indirect standardization is preferable when explaining spatial differences. Rates per class of age (per class of the standardization variable in the general case) are then used. They are independent from receiving units of aggregation, which are applied identically to all of these objects. Indirect standardization then operates as a global model. Spatial distribution of differences between values yielded by the model and observed values can be studied later.

SMR (Standardized Mortality or Morbidity Ratio) is used in epidemiology after indirect standardization (generally on age, which is the most common risk factor in epidemiology). It is the ratio between the observed number of patients O and the expected number of patients E (calculated depending on general rates). In epidemiology, it is generally used on the set of individuals, with a significance test that involves testing the hypothesis $H0 : SMR = 1$ (no difference between observed and expected values), using the Breslow–Day test [BRE 87, BOU 93] and Samuel test [SAM 91] or simulation.

For each object, the hypothesis of an SMR differing from 1 can be evaluated by a test that involves calculating the value X, which follows an χ^2 law with 1 degree of freedom:

$$X = \frac{|N_{Observed} - N_{Expected} - 0.5|^2}{N_{Expected}} \text{ (Breslow–Day test)}$$

When $N_{Expected}$ is small, the approximation of Poisson's law by the normal law is less satisfactory, and the expression must be replaced by:

$$X = 9 \times N_{Observed} \times \left(1 - \frac{1}{9 \times N_{Observed}} - \sqrt[3]{\frac{N_{AExpected}}{N_{Observed}}}\right)^2 \text{ (Samuel test)}$$

The confidence interval $[SMR_{Min}, SMR_{Max}]$ of SMR is given by:

$$SMR_{Min} = \frac{N_{Observed}}{N_{Expected}} \times \left(1 - \frac{1}{9 \times N_{Observed}} - \frac{Z_{\frac{\alpha}{2}}}{3 \times \sqrt{N_{Observed}}}\right)^3$$

and

$$SMR_{Max} = \frac{N_{Observed} + 1}{N_{Expected}} \times \left(1 - \frac{1}{9 \times (N_{Observed} + 1)} - \frac{Z_{\frac{\alpha}{2}}}{3 \times \sqrt{N_{Observed} + 1}}\right)^3$$

The probability of various observed values can also be calculated by simulation. Monte Carlo method is used to simulate the distribution of (expected) cases in the set of objects, depending on the population of each object and on the probability of each individual being positive. For each object, this leads to obtaining the characteristics of the (quasi-normal) distribution of simulated values (mean and standard deviation), which makes it possible to assess the probability of the observed situation.

Box 6.4. *Breslow–Day test and Samuel test*

In the case of localized data aggregated into geographical units, the SMR can be calculated one geographical unit at a time, and will be used to analyze and compare differences between units.

SavGIS	The significance of SMR can be calculated with the menu: Calculate → Statistical calculations → Significance (SMR)

Table 6.4. *In practice*

6.2.2.3. *Study of the statistical distribution of SMRs*

QUESTION.– Has a given unit an "abnormal" value?

As noted in Chapter 4, the reader of a map focuses on the geographical units with extreme values, but these units are not necessarily abnormal. The question concerns only one unit each time, independently of the others. It must be verified if a unit appearing as "abnormal" compared to the others has a value that effectively allows rejecting its null hypothesis $H0 : SMR = 1$.

A Breslow–Day test is used to test the null hypothesis $H0 : SMR_i = 1$ for unit i, having as alternative hypothesis $Ha : SMR_i \neq 1$ in the double-tailed case or $Ha : SMR_i > 1$ $(or\ SMR_i < 1)$ in the one-tailed case. If individual data are available before aggregation, classic tests for mean or proportion comparison can also be used.

QUESTION.– Are there geographical units whose value is "abnormally" high or low? What are the units that present these "abnormal" values?

This question differs from the previous one because the analysis and representation of a set of aggregation units implicitly and visually addresses a more global alternative hypothesis, of the type "there is *at least* one abnormal unit in the set of units". With respect to the first question, we are faced with a multiple test problem: the more geographical units, the higher the chance that one of these units rejects its individual null hypothesis by chance (rejection is a function of the chosen risk of error). This approach does not seem natural, since when analyzing a map, we are inclined to find an *explanation* for the observed differences, no matter how small, rather than admitting their occurrence may be due to chance. This is especially valid for extreme values that are easily detectable on the map, even though they are not necessarily very "different" from the set of values (Figure 6.2).

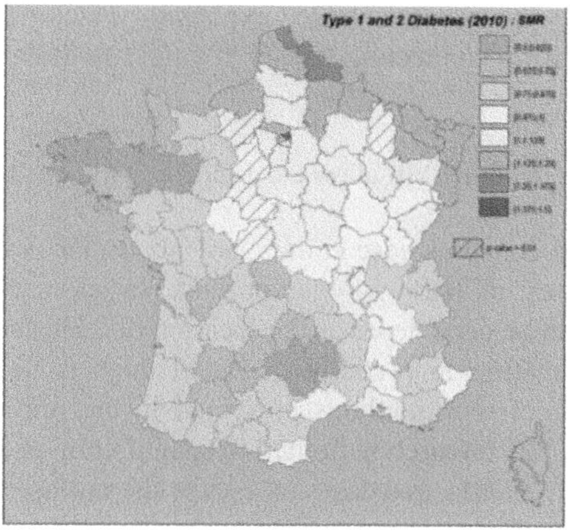

Figure 6.2. *Diabetes in metropolitan France (2010): value of SMR by department and indication of its significance (the unhatched zones individually have an SMR significantly different from 1, with a risk of error of 1%). (Indirect) age standardization has been realized with 2010 rates calculated on the set of districts (three classes of age: 0–45, 45–75, >75). For a color version of this figure, see www.iste.co.uk/souris/epidemiology.zip*

Therefore, the analysis concerns the global distribution of the values of the units. The question corresponds to highlighting the aggregation units that have "abnormal" values with respect to a random distribution or to a given theoretical distribution. As discussed previously, a Breslow–Day test is used for each unit, but the risk of error is reduced in order to address the multiple test problem.

It is also possible to compare units two by two (or each unit to a "mean" unit), but it should be noted that a direct comparison of SMRs may be misleading, in the sense that it does not reflect the comparison of incidence rates in some particular cases (if there is a combination of two phenomena: very different rates and very different structures of the standardization variable).

SavGIS	Tests for the analysis of distribution and variance between a set of geographical units are available with the menus: Stat → Statistical tests → Distribution and Stat → Statistical tests → Variance analysis

Table 6.5. *In practice*

When the approach is explanatory, as will be noted below, the objective is to explain the weakest or the strongest values of the distribution, even though they are not statistically abnormal and could therefore be generated by chance. Then, the analysis brings in one or several supposed explanatory factors, in order to show the existence of a relationship connecting the set of values to these factors (and not only the extreme values).

QUESTION.– Does the analysis of SMR allow the detection of clusters?

When aggregation concerns case counts, an "abnormal" value represents a lack or an excess of cases in the aggregation unit. In case of excess (significant SMR > 1), the unit can be considered as the location of a cluster of cases. The search for clusters by aggregation in predefined spatial units is nevertheless much less powerful than the direct scan methods presented in the previous chapter. Indeed, with aggregated data, this research is limited to spatial units of aggregation – the size and shape of which are fixed – and does not escape the multiple test problem.

Similar to the scan-based search for clusters, the multiple test problem can be bypassed by considering, for example, the value $X = |SMR - 1|$ as an index, by preserving only the unit with the highest X, and testing only this unit. The analogy with the Kulldorff method for cluster search is obvious. The geographical unit corresponds to a research window. The size of the window does not vary here, and the position of the center only varies among the centroids of the units. The index used also differs, as it does not directly take into account individual probabilities. A scan-based search for clusters can therefore yield a result that is different from the unit with the farthest from 1 SMR.

QUESTION.– Does the distribution of values of the units present an "abnormal" variance, such as an overdispersion?

In addition to highlighting extreme values, the characteristics of the distribution of values of geographical units can be analyzed (generally, after standardization or on a residue). The objective is to analyze the entire distribution, and not a particular unit. The aim is to find out whether or not the distribution of values (in general, SMR) is in accordance with a given distribution of the variable in the set of units, such as a random distribution or a more specific distribution (for example, a distribution presenting a non-null autocorrelation, corresponding to a phenomenon of contagion by proximity). The number E_i of expected values in each unit will be calculated depending on this theoretical distribution.

The question raised concerns the heterogeneity of risk over the entire studied region. *Overdispersion* (conversely, *underdispersion*) is defined as a variance of the observed values O_i significantly above (or below) the theoretical variance resulting from the model used for the estimation of expected values E_i [GAL 04, DEA 92]. Aggregation units are *a priori* considered independent.

A simple $\chi 2$ test with $n-1$ degrees of freedom can be used to compare the distribution of observed values and the distribution of expected values ($\chi 2 = \sum_i \frac{(O_i - E_i)^2}{E_i}$); however, this test does not have a very high power, and is not suitable for low counts per units (which is often the case when there are many geographical units) and therefore a significant variance of SMR. Therefore, Potthoff–Whittinghill test is used [POT 96a, POT 96b].

Potthoff and Whittinghill have described an overdispersion test that is more powerful than a simple $\chi 2$ test. The alternative hypothesis is no longer elaborated as the hypothesis complementary to H0 (Ha: there is a unit with Oi ≠ Ei), but with an additional hypothesis on risk distribution (for example, probabilities in the units result from a beta distribution or from a negative binomial distribution).

In the case of a beta distribution, the index (test statistic) used in a Potthoff–Whittinghill test is:

$$\mathrm{S} = \mathrm{E} \sum_i \frac{O_i(Oi-1)}{E_i}.$$

The statistical distribution of S is quasi-normal, of mean $O(O-1)$, and variance $2(m-1)O(O-1)$.

When the observed risk variance is above their mean (overdispersion with respect to a Poisson model), a hypothesis of a negative binomial distribution of probabilities is formulated, with $(Variance/Mean) = 1 + \beta$. β is given by:

$\beta = \frac{A}{C}, with\ A = 1/2\left[\sum_{i=1}^{m}\left(\frac{O_i*(O_i-1)}{E_i}\right) - (O-1)\right]$ and $C = \frac{1}{2} * \frac{(m-1)*(O-1)}{O}$.

The p-value of the observed value of β is estimated by simulation, randomly distributing the total number of cases in the units depending on probabilities expected in each unit.

Box 6.5. *Potthoff–Whittinghill test*

Overdispersion analysis is not, strictly speaking, spatial analysis, as the spatial relationships between geographical units are not taken into account (the units are considered independent). Nevertheless, dispersion can be the result of an actual difference in probability for each geographical unit (that can be explained, for example, by relationships with the environment), or of a correlation between objects in a unit (for example, a proximity relationship, such as contagion). Risk heterogeneity may be observed in the absence of a specific spatial pattern.

QUESTION.– How are values or counts distributed within units? How can inequalities between units be measured?

Distribution analysis concerns the concentration of values (modality counts or sums) in the units (for example, 20% of units contain 80% of cases). Inequality indices are used for this purpose, particularly the Gini coefficient [GIN 21], Theil index [THE 71] or the localization quotient [ISA 60] (that corresponds to SMR in case of a Boolean variable). These indices are widely used in geography or economy. To test the index significance, its distribution must be evaluated by the Monte Carlo simulation.

Let N be the number of units, A_i the count of unit i, O_i the value of unit i, O the sum of all units and A the total count. Units are sorted in ascending order of their value $(O_{i+1} \geq O_i)$.

The Gini coefficient is a measure of distribution dispersion. It ranges from 0 (uniform distribution) to 1 (maximum inequality):

$$G = \frac{2\sum_{i=1}^{N} i\, O_i}{N\sum_{i=1}^{N} O_i} - \frac{N+1}{N}.$$

The Theil inequality index with respect to the variable is:

$$T = \ln\left(\frac{A}{O}\right) - \frac{\sum_{i=1}^{N} O_i \ln\left(\frac{A_i}{O_i}\right)}{O}.$$

The inequality index with respect to the population is obtained by reversing A and O.

A localization quotient is a two-way classification table: one for the units, and the other one for various modalities of the variable. Each case contains the ratio between observed count and expected count.

Box 6.6. *Gini coefficient, Theil index and localization quotient*

6.2.2.4. *Low counts and low probabilities*

Sometimes, the confidence interval of the SMR of geographical units is wide and does not allow for differentiation between the units that present no cases and the units that present one (or two) cases. This is particularly the case when counts or probabilities are very low, and when the distribution of cases in the units cannot be distinguished from a random distribution (generally modeled by a Poisson distribution). Assessing the probability of the phenomenon in each unit by means of unit rate should be avoided, owing to very poor results (too wide of a confidence interval). Here the focus is rather on studying "abnormal" units (as previously defined) or on searching for a necessary condition for the occurrence of an event, with classic tests of statistical comparison. The condition may be an environmental variable (assuming an environmental risk factor) or a unit characteristic variable (such as population density or frequency of contacts, assuming a risk factor linked to the proximity of individuals). If the number of cases is small, it is preferable to analyze non-aggregated data and to resort to an analysis of events themselves rather than to an analysis of unit rate.

When aggregation is imposed, the susceptible population in each group (N_k, with the previous notations) is sometimes unknown, and consequently cannot be used to calculate ratios (incidence, prevalence, etc.). Nevertheless, under certain conditions (very low probabilities, small sizes of population), it is possible to make the approximation of uniform distribution of population N among groups G_k, which amounts to dealing with the counts of positive values O_k, instead of ratios O_k/N_k.

Indeed, suppose the probability P is identical in each group. The number of expected cases in each group is N_kP, where N_k is the count of group k. When all N_kP are small (either because k is large with respect to NP, P is small or N_k are small), most groups present no cases, even for probabilities P'_k that are very different one from another. For example, for an individual probability of 0.001, the probability of obtaining at least one case will be 0.09 in a village of 100 inhabitants, and 0.18 in a village of 200 inhabitants. Despite their difference in population size, it is very probable that neither of these villages will present any cases. Even though there is a large variation in the population size from one village to another, its influence on global risk is weak. Obviously, the most densely populated villages have more cases overall.

In summary, the analysis of individual data aggregated at the level of G_k can use:

– O_k/N_k ratio or a variable allowing the comparison of the observed value of a probability to its theoretical value calculated according to the principle of maximum likelihood from individual probabilities (O_i/E_i). It is the natural variable, with the remarks stated above on the variance of probability assessment;

– the number of positives O_k. It is the case of unknown or small N_k or of very low individual probabilities;

– one Boolean variable: 0 if $O_k = 0$, and 1 if $O_k > 0$. This type of variable is used similarly to the previous case, and when differences in probabilities of individuals must be ignored, if they exist. The Boolean variable allows the characterization of the aggregation unit status, independently of its population of susceptible individuals.

Box 6.7. *Low counts or low probabilities*

6.2.2.5. *Search for environmental correlations*

QUESTION.– Is it possible to highlight correlations between aggregated values and environmental variables characterizing the geographical units? Is it possible to find a statistical model that explains the values observed from the variables of the geographical units?

This approach is *explanatory*: the objective is to explain the lower or higher values of the distribution, even though they are not statistically abnormal and could therefore be generated by chance. One or several assumed explanatory factors should then be brought in, and the existence of a relationship connecting the aggregated values to explanatory factors should be identified, which explains the values (not only the extreme values). In this approach, the influence of known risk factors should first be eliminated (by standardization or modeling) in order to analyze only the residue of disparities due to assumed explanatory factors.

Finally, the statistical analysis of the aggregation units concerns the search for correlations between the attributes of these units. It is thus very common to use environmental values obtained from layers of information concerning the environment, and also calculated by aggregation (using particularly the possibilities of spatial join of the geographic information systems). For example, it is common to use remote sensing data to estimate environmental parameters in geographical units, and to use these parameters in statistical analyses (correlations, multivariate analyses, regressions, etc.).

The objects of these correlation calculations are the geographical units used for the aggregation of individuals, and not the individuals themselves. As the data that characterize them are obtained by aggregations, they generally represent means or rates, their validation must be discussed, with the restrictions that have already been indicated. The conclusions of these analyses on the units should not be inferred to individuals. Moreover, they do not consider direct spatial relationships between individuals except for the belonging to the same geographical unit.

Finally, known risk factors should be eliminated by standardization or modeling of covariance in order to avoid any false interpretation.

EXAMPLE.– It is known that the age of puberty for girls partly depends on genetic, ethnic and socio-economic factors. Some endocrine disruptors present in the environment are also presumed to have an influence. To highlight this specific relationship during a survey, all factors should be considered. If the spatial study is conducted on aggregated data, the influence of individual factors should be eliminated (by standardization or modeling) before analyzing the disparities in the spatial distribution of results and before searching for correlations with environmental factors.

SavGIS	Correlation analyses are available with the menu: Stat → Correlations and Stat → Covariances and ACP Statistical models are available with the menus Calculate → Statistical calculation → ACP and Calculate → Statistical calculation → modeling

Table 6.6. *In practice*

6.2.3. *Spatial analysis of aggregation*

Localization brings additional information on the distribution of unit values. Spatial analysis focuses on highlighting the characteristics in the spatial distribution of values of the units: overdispersion, autocorrelation, hot spots, clusters, trends and gradients. The analysis no longer concerns the individual values of the units, unit by unit, but the set of units taken as a whole. The individually non-significant values can therefore be considered, since here significance is sought for the spatial distribution of the whole set of units or of a group of units. The analysis concerns spatial relationships between values and their distribution but never the spatial distribution of units themselves, since these are chosen independently of the studied phenomenon.

QUESTION.– Do the studied geographical units have the same internal variability? Is the classification by aggregation in the chosen geographical division relevant for the spatial differentiation of the phenomenon? Is the geographical division used adapted to the problem at hand? Is the spatial scale of aggregation adequate for the analysis of the phenomenon?

The analysis refers to spatial aggregation quality and to probability assessment. Variability between units (inter-class inertia) and variability inside each unit (intra-class inertia) are studied. Statistical indices allow the measurement of these characteristics. The most frequently used indices are TAI (Tabular Accuracy Index) and GVF (Goodness of Variance Fit).

Given n values x_i in an Euclidean space with a distance d, and a partition in q classes of center of gravity g_j, and inertia $I_j = \sum_{i=1}^{n_j} d^2(x_i, g_j)$,

Intra-class inertia is: $I_{intra} = \sum_{j=1}^{q} I_j$.

And inter-class inertia is: $I_{inter} = \sum_{j=1}^{q} d^2(g_j, g)$.

where g is the center of gravity of the set of values. Total inertia of the cloud of points is the sum of inter- and intra-class inertias. Two partitions of the same number of classes can thus be compared. Two indices are used, namely TAI (Tabular Accuracy Index) and GVF (Goodness of Variance Fit):

$$TAI = \frac{\sum_{k=1}^{q}(\sum_{j=1}^{n_q} |x'_j - \frac{1}{n_k}(\sum_{\iota=1}^{n_k} x'_\iota)|)}{\sum_{i=1}^{n} |x_i - \bar{x}|}$$

$$GVF = \frac{\sum_{k=1}^{q}(\sum_{j=1}^{n_q}(x'_j - \frac{1}{n_k}(\sum_{\iota=1}^{n_k} x'_\iota))^2)}{\sum_{i=1}^{n}(x_i - \bar{x})^2}$$

where x'_j is the jth element of group k. GVF ranges from 0 (maximum internal variance in classes) to 1 (no internal variance in classes).

The ratio $R = {}^{I_{intra}}/_{I_{inter}}$ is also used in order to promote the partition that minimizes it.

Box 6.8. *Intra-class inertia and inter-class inertia*

SavGIS	TAI and GVF are available with the menu: Quest → Geoaggregation and all the menus of Classify item

Table 6.7. *In practice*

As already underlined, aggregation may be imposed (when data are already aggregated, for logistic, administrative or ethical reasons) or chosen, when the precise location of individuals or events is available. In the latter case, aggregation itself can be used for spatial analysis purposes choosing the size and shape of the receiving units. For example, a cell division can be created (square or hexagonal unit cells) and the size of these cells can vary in order to determine the optimal size (maximization of GVF or minimization of the ratio *R* of intra- and inter-class inertias).

Quadrant spatial analysis involves spatial aggregations in cells that divide space, and varying the size of the cell (Figure 6.3). Statistical analysis allows at each level the analysis of group homogeneity, comparing the distribution of counts or values to a random distribution, extracting the "abnormal" cells and analyzing global spatial trends. Since an increase in the number of aggregation units increases variability, (because it reduces the probability of occurrence of the phenomenon in each unit), it is interesting to find the finest scale allowing the detection of a spatial abnormality, such as a cluster (which involves high SMR in the cell).

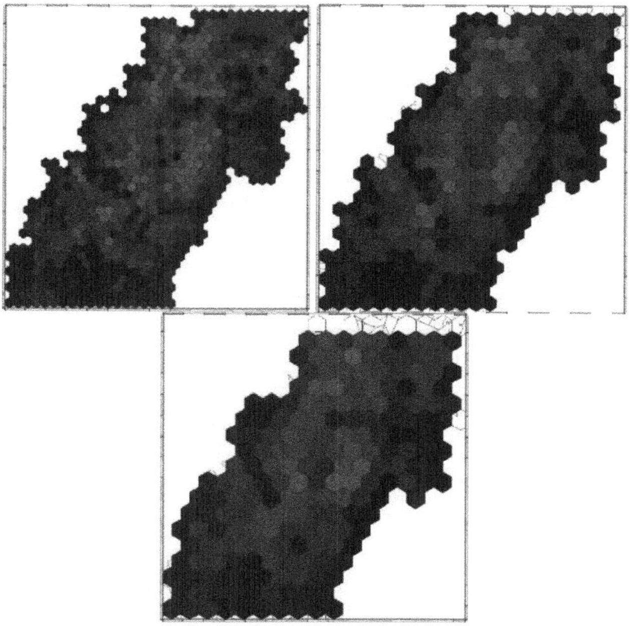

Figure 6.3. *Aggregation in unit cells of various sizes for the search of a better scale of analysis (population density, Quito center, Ecuador)*

QUESTION.– Does the phenomenon observed at the level of geographical units of aggregation have specific spatial characteristics?

The techniques used for the study of autocorrelation, detection of clusters, trends and gradients are those described in Chapter 5. Unit localization is schematically represented by its centroid. The methods can also use topological or metric relationships between units (adjacency, distance).

The measure of autocorrelation between geographical units generally uses a contiguity matrix (1 if two units are adjacent, 0 otherwise) rather than a weights matrix depending on the distance between centroids. Several adjacency orders can also be used (for example, second order: adjacent to adjacent) with different weights and a correction to avoid the over-representation of second-order units, which are more numerous. When calculated on data resulting from another division (as an exposure factor), a significant autocorrelation can be generated by improper adequacy between geographical units of aggregation and spatial distribution of the risk factor.

6.2.4. *Spatial belonging*

In contrast to aggregation, spatial belongings allow the creation of new attributes for the individuals, using their belonging or their distance to geographical units or, more generally, to spatial objects.

Statistical and spatial analyses at the individual level are obviously more powerful than the analyses of geographical units of aggregation, as they allow the consideration of the agents' individual characteristics, as well as the spatial relationships between agents. Spatial belonging and spatial join operations can be easily implemented in geographic information systems, and GIS are frequently used to add environmental variables by operations using localization (spatial join, search within a radius, buffer, vicinity, etc.). Thus, localization allows the addition of many context attributes to individuals, who are then analyzed and modeled by classic statistical methods.

The quality of the "spatial join" at the level of individuals must nevertheless be questioned, particularly with respect to spatial resolution and hence to the spatial accuracy of environmental data, or with respect to the coherence between the location of health events and the location of individuals: does the location of the individual coincide with the location of the health event? Is it sufficiently accurate? Are the environmental parameters obtained by belonging actually the parameters at the moment of the health event? Is the spatial accuracy of the available environmental data compatible with the studied phenomenon?

6.3. Statistical modeling of spatial data

The purpose of statistical modeling is to find a mathematical expression that allows the estimation of the value of an attribute observed from a set of other attributes, referred to as explanatory. The analyzed objects are "individuals" in the statistical sense, but modeling applies to geographical units whose variable to be explained results from a process of aggregation of agents, as well as to agents themselves. Explanatory attributes can be classic descriptive attributes or attributes resulting from aggregation, spatial belonging or spatial join (for environmental data in geographical units, for example). In the case of geographical units, the value to be explained by the model is a mean, a sum, a count or a ratio, while in the case of an analysis of agents, the value to be explained corresponds to a probability or to a Boolean value (for example, sick/not sick).

6.3.1. *Statistical correlations and spatial relationships*

As already underlined several times, a health phenomenon can be governed at the level of an individual simultaneously by individual factors and by factors stemming from (notably spatial) relationships with other individuals. When the studied process depends on proximity relationships between individuals, the latter cannot be considered as independent for a statistical analysis concerning another characteristic, such as a search for an environmental correlation, a regression or a statistical test. When individuals are not independent and identically distributed (i.i.d.) with respect to the studied statistical relationship, the statistical result is biased: a correlation can be artificially reinforced, a statistical test can erroneously reject its null hypothesis, as the non-independence of individuals can reduce the p-value of the observed situation. In order to correct the bias introduced by the spatial non-independence of individuals, it is possible to weigh the individuals depending on their spatial dependence or to include these spatial dependences in the statistical model or in the evaluation of the distribution under the null hypothesis (for example, using a model that creates situations involving clusters, such as a Cox process).

EXAMPLE.– The example of a contagious disease facilitates the understanding of this problem. The observed cases show a strong clustering due to contagion, and therefore cases are mainly found in the types of habitat that spatially correspond to this cluster. A statistical study hence shows a strong correlation between case and type of habitat. In fact, this correlation may stem uniquely from the clustering of cases due to the contagious character without being linked to whatever relationship exists between the disease and the characteristics of the habitat of individuals; the fact that cases are clustered for disease transmission reasons and consequently are not independent and identically distributed, highlights the statistical relationship between disease and habitat (but this relationship can also exist elsewhere, since transmission is related to a vector that is itself correlated with the type of habitat).

Considering time is also essential, as it prevents one from considering individuals that have no link to be linked by spatial relationships (they are neighbors, but they do not belong to the same sequence, and can therefore be considered independent with respect to the studied phenomenon).

6.3.2. *Statistical modeling*

QUESTION.– Find a mathematical expression that makes it possible to model an individual health risk from individual and environmental characteristics, including factors related to relationships with other individuals.

The general objective of statistical modeling in epidemiology is to explain or model an individual risk of disease Z from individual and contextual risk factors. The objective is to find a functional expression of the type $Z = f(x1, x2, \ldots, xn, \varepsilon)$, which allows the modeling for all the individuals of the studied value Z as a function of various risk factors $x1, x2, \ldots, xn$, and to analyze its variance and residue (the mean and the distribution of differences between the model and the observed values). The classic approach is to use linear and multilinear regressions, generalized multilinear regressions and Poisson regressions, depending on the observed or assumed distribution of Z. For example, a linear model has the following form:

$$z_j = \beta_0 + \sum x_{jk}\beta_k + \varepsilon_j$$

where x_{jk} are the values of risk factors for individual j, β_k the coefficients for the linear combination and ε_j a random error. Using matrices:

$$Z = X\beta + \varepsilon$$

To ensure the quality of the model, all errors ε_j are required to have a mean of 0, identical variance and no inter-correlation.

Model parameters are determined from observed data by classic data analysis methods (least squares, maximum likelihood). Classification methods based on machine-learning techniques (for example, *Boosted Regression Trees* [ELI 08] or the *Analytic Hierarchy Process* [SAA 08]) can also be used.

The simplest method involves finding a linear relationship between one variable and the other ($z = ax + b$). When the problem is multivariate, the expression is $z = \beta_0 + \beta_1 x1 + \beta_2 x2 + \cdots + \beta_n xn$.

The typical form of generalized multilinear regressions is $f(z) = \beta_0 + \beta_1 x1 + \beta_2 x2 + \cdots + \beta_n xn$, where f is a function. An example is logistic regression, which aims to model $ln(p/1-p)$ by a linear expression from an observed probability p (obtained by estimation depending on a number of cases and a susceptible population). When the variable to be explained is a count and follows approximately a Poisson distribution ($P(z = k) = e^{-\lambda \frac{\lambda^k}{k!}}$, particularly for low values of k), a direct linear relationship with the explanatory variables is often not adapted, therefore a Poisson regression or a log-linear regression will be used to evaluate the mean of the searched Poisson distribution that corresponds to $E[z]$. The form of the regression is $ln(\lambda) = \beta_0 + \beta_1 x1 + \beta_2 x2 + \cdots + \beta_n xn$. The model can be applied to a count (or to the numerator of a ratio, for example, a standardized incidence).

The model can also involve linear combinations of nonlinear functions (power functions, trigonometric functions, etc.) or products of variables.

Coefficients β_k are determined from observed data by classic data analysis methods (minimization of mean square error, maximum likelihood) or machine-learning methods (regression trees, AHP).

There are many "potential" risk factors. Those whose contribution to the model is small can be eliminated individually, analyzing the values of agents with respect to each factor (correlation coefficients, ANOVA or the Kruskal–Wallis test), or globally, by gradual elimination of those with the smallest coefficient in the regression until a given threshold is reached. Analyses into main components or discriminant analyses make it possible to solve the problems of dependence between risk factors and reduce the number of factors involved in the model, by eliminating the components whose correlation with the variable to be explained is too weak or whose variance is too weak.

Box 6.9. *Statistical modeling*

Statistical modeling allows the consideration of the influence of environmental risk factors and context variables. These environmental and contextual variables are generally obtained by a belonging relationship or spatial calculation.

6.3.3. *Spatial models*

When the phenomenon is localized and its location matters, these models are *a priori* incomplete, as they consider neither the relationships between individuals, nor the autocorrelation of risk factors that were not considered in the model. The study of the model residue contributes to verifying if its spatial distribution is random: if this is not the case, this means that a geographical risk factor or a process involving relationships or spatial dependences between individuals has not been considered in the model. On the contrary, when risk factors are localized and present a spatial autocorrelation, this additional information allows an increased accuracy of the model, as for the evaluation of Z in a point P, it is possible to bring in the values of risk factors in the vicinity of a point P.

Therefore, the main idea is to consider the values of neighboring individuals in the estimation of an individual [ANS 88]. Many models, referred to as spatial autoregressive models, have thus been developed. They rely on a very simple idea: in the model one adds a component that depends on the neighboring objects to "apprehend" that which generates the spatial dependence of the variable to be explained. The simplest idea would be to directly model the value z'_j resulting from the mean or the weighted sum of the values z_{i_k} of the objects in the vicinity of object j.

This is not a matter of adapting the model to the place. The objective is the same: finding a unique model, applicable to all individuals irrespective of their position. A phenomenon is assumed to be governed by a unique law that does not depend on place, but only on the value of risk factors and on the presence of other individuals in the neighborhood: the phenomenon is still assumed to be stationary.

There are two main types of spatial models: on the one hand, models that attempt to directly model the spatial autocorrelation of Z, relying on the result of the process and ignoring the source of this autocorrelation, and on the other hand, those relying on the residue, which attempt to model the spatial structure of this residue independently of the values of coefficients β_k.

In order to model the spatial autocorrelation of Z, an additional term in the form of a function f of other agents is introduced in the model. This can be a very general function whose value depends on the value of the other individuals, particularly neighbors:

$$z_j = \sum_k x_{jk}\beta_k + f(z_1, z_2, \dots) + \varepsilon_j \ (Z = X\beta + f + \varepsilon)$$

where z_j is the value of the model at point P_j, and β_k are the coefficients of the linear regression corresponding to explanatory variables. The addition of f to the model operates as a spatial smoothing, as it finally comes down to a classic modeling of the values $z_j - f(z_1, z_2, \dots)$.

These various models, called autoregressive, depend on the definition of f. The most frequently used models are autoregressive models, simultaneous autoregressive models (SAR) and conditional autoregressive models (CAR) [DRO 06, DOR 07, FOT 09], whose function f is a linear function of neighbors. What differentiates these models is not the form of f, which is still linear, but the object to be modeled by f. For example, for autoregressive models, the following can be written:

$$z_j = \sum_k x_{jk}\beta_k + \rho \sum_{i,i\neq j} w_{ij} z_i + \varepsilon_j \ (Z = X\beta + \rho WZ + \varepsilon)$$

where w_{ij} is a spatial weight that depends on the distance between the points i and j, and ρ is a factor allowing the parameterization of force and the nature (attraction or repulsion) of the spatial dependence in the model. Spatial weights may correspond to a relative weighting between neighbors: for all the individuals j, the sum of weights of all the neighbors $\sum_i w_{ij}$ must then be equal to 1, and $\sum w_{ij}y_i$ is a weighted mean. If all the weights are equal, this comes down to adding to the model the mean of neighboring values. They can also have an absolute influence. In this case, the more neighbors close to z_j, the higher the value of $\sum w_{ij}y_i$. The model is added a weighted sum of the values of neighbors, and not a weighted mean.

SAR models (simultaneous autoregressive models) may various forms:

$Z = X\beta + \rho WZ + \varepsilon$: the model only considers the autocorrelation of the values to be explained.

$Z = X\beta + \rho WZ + \gamma WX + \varepsilon$: the model explicitly and separately considers the autocorrelation of the variable to be explained and that of the explanatory variables, each component having a distinct parameter.

$Z - X\beta = \rho W(Z - X\beta) + \mu$: the model only concerns the autocorrelation of residue ε, and the residue μ of the final model must be spatially random. Spatial weight only applies to the part of value z_j that cannot be explained by the linear combination $X\beta$ of explanatory variables.

CAR models (conditional autoregressive models) only consider the autocorrelation of the local residue $(Z - X\beta)$, with a weighted mean of the residues of neighbors. The weights matrix W must be the symmetric positive definite, which means that the interactions between two agents must be symmetric (relationships between two agents i and j are not connected to the order of the pair i, j). It must be column normalized ($\sum_i w_{ij} = 1, \forall j$).

The library **spdep** of the R software program contains commands available for the use of these statistical models (as well as many other spatial analysis functions [BIV 17]). Other software programs also allow the implementation of autoregressive models, such as GeoDA, Matlab, SAM, SpaceStat and S-plus.

In epidemiology, the choice of model depends on the type of variable to be modeled (individual risk or aggregated value) and on the hypotheses formulated on the process. If the unique objective is to consider the unexplained spatial dependence without distinguishing the autocorrelation of ignored risk factors and spatial relationships between individuals, a simple autoregressive model should be chosen. It is often the case for a value to be explained that corresponds to values aggregated by geographical units in which spatial relationships between individuals are lost, or when an aggregation made in units that are too small with respect to the studied phenomenon involves a weak inter-unit variance and a strong autocorrelation between neighboring units.

All of these models rely on linear models. Spatial structures and relationships often induce nonlinear relationships, and a non-Gaussian distribution of the variable to be explained or of the residue is then noted. Linear models are therefore not necessarily very adapted to considering variables that present spatial dependences. GAM models (generalized additive models [HAS 90]) and STAR models (structured additive regression models [KNE 05]) have been developed to model the nonlinear effects of variables in the non-Gaussian case or for regressions with non-Gaussian residue.

> STAR models are the extensions of GAM models which allow the integration in a common model of local spatial effects, nonlinear effects of risk factors and linear effects or usual fixed effects.

Box 6.10. *Structured additive regression (STAR) model*

6.3.4. *Spatial heterogeneity of parameters*

QUESTION.– Find a mathematical expression that allows the modeling of a health risk from individual and context characteristics without assuming the phenomenon to be stationary.

Spatial instability of the parameters of a regression is a manifestation of a spatial heterogeneity of the studied phenomenon, which corresponds in fact to one or more processes that do not have the same characteristics throughout the studied domain. The studied process is therefore not assumed to be stationary.

It would be preferable to differentiate the regions and build several models that are assumed stationary, thus allowing the consideration of particular characteristics of each region, with a model adapted to each region. However, the problem can also be addressed in a continuous manner, by trying to adapt the model to the local context or to directly translate the variations of the variable to be explained in a model whose parameters depend on localization.

6.3.4.1. *Geographically weighted regression (GWR)*

The process to be modeled, still in the form of a multilinear model, is no longer assumed stationary, unlike the above-described autoregressive models. Rather than adding a supplementary term to the model, the coefficients β of the model can vary depending on localization, to locally adapt the model to the observed local spatial variations of Z. The model unfolds into as many local models, having for each individual j:

$$z_j = \sum_k x_{jk}\beta_{jk} + \sum_{i,i\neq j} w_{ij}\left(\sum_k x_{ik}\beta_{jk}\right) + \varepsilon_j \; (Z = X\beta_j + WX\beta_j + \varepsilon)$$

W is still a spatial weights matrix, obtained as in the previous chapters.

This is how GWR (geographically weighted regression) models [BRU 98, FOT 02, YU 10] have been developed. The calculation of coefficients β_j at point j uses the values of individual j and the values of other individuals i weighted by a spatial weight that depends on their distance to individual j. The method can be applied to any type of model, particularly to generalized linear models, as long as the equations allowing the determination of the coefficients can be solved. GWLR (geographically weighted logistic regression) models are such an example.

GWR models allow the adaptation of the regression to local variations, by locally modifying the regression coefficients. This comes down to elaborating an equal number of models and individuals; the approach does not allow the elaboration of a global model (explanatory or predictive), and is therefore in contradiction with the objective of modeling the process in its entirety. The model allows the integration in the expression of the statistical model for individual j of the local spatial autocorrelation of risk factors. On the contrary, unlike SAR and CAR models, GWR models do not allow the consideration of spatial relationships between neighboring individuals that are not linked to individual risk factors included in the model.

The main interest of weighted regression is to allow the visualization of the spatial variations of the coefficients of each risk factor, and therefore the important spatial variations of each factor in the type of model used. The spatial variation of coefficients may allow the definition of more homogeneous spatial sub-domains, corresponding to different statistical models. Spatial analysis of the variations of coefficients is particularly interesting for the detection of places corresponding to the high relative influence of a specific factor, and particularly for risk factors allowing a public health action aimed at risk reduction.

> The classic regression model is modified to allow the local estimation in a point P of the regression parameters, considering the observations in a given radius and assigning a weight to observations depending on their distance to point P. The weight is calculated classically, using a function of distance, such as kernel estimation.

Box 6.11. *GWR (geographically weighted regression) models*

Several software programs can be used for the implementation of spatial-weighted regressions: GWR4 package [NAK 15], ArcGIS, R [BIV 08], SavGIS, SAS and STATA.

SavGIS	GWR modeling is available with the menu: Calculate → Statistical calculation → GWR model
ArcGIS	GWR modeling is available with the menu: ArcToolbox → Spatial Statistics Tools → Modeling Spatial Relationships → GWR
QGIS	GWR modeling is available with the menu: Processing Toolbox → SAGA → Geostatistics → GWR

Table 6.8. *In practice*

6.3.4.2. *Spline functions*

The idea is to divide the domain of variation of an explanatory variable into segments, and to search each segment in order to model the variable to be explained by a polynomial function of nth degree, ensuring the continuity of the polynomial at the knots between segments. An *nth degree spline* (piecewise polynomial function of one or more variables) is thus defined.

This approach corresponds to a *discretization* of one or more explanatory variables, which corresponds itself to a discretization of space into spatial subgroups. The segmentation of the explanatory variable(s) is done *a priori*, and should ideally correspond to spatial sub-groups that reflect structural changes that are to be expected or have already been highlighted during the cartographic or spatial analysis.

EXAMPLE.– If a phenomenon is assumed to vary as a function of the distance to a point (for example, density with respect to a center), dividing this distance into segments allows the definition of spatial sub-groups (rings around the center), each sub-group having a model in the form of a polynomial, while ensuring the continuity of the derivatives of polynomials at the knots between segments.

6.3.4.3. *Trend surface analysis*

The approach is no longer discrete but continuous, and the objective is to directly regress the variable to be explained by a single polynomial function of geographical coordinates. The polynomial coefficients are assessed in a classical manner, by the least squares method. This *trend surface* allows the description of significant trends that characterize the regression surface. As it has already been mentioned in the mapping section, since the value of the polynomial is a function of coordinates, it can be readily calculated and represented on a map.

This type of model nevertheless has many limits: it does not allow the elimination of the spatial autocorrelation of the residue, and the various explanatory variables (powers and products of coordinates) are strongly inter-correlated. The choice of the degree of the polynomial is also difficult to justify. It generally involves an ascending approach, starting with lower degrees. Finally, and above all, a trend surface has no explanatory character in terms of risk factors, since intrinsic localization is generally just a confounding factor of little interest for the epidemiologist.

SavGIS	Trend surface calculation is available with the menu: Calculate → Statistical calculation → Multivariate modeling

Table 6.9. *In practice*

6.3.4.4. *Kriging regression*

The spatial models that have been described above use weighted sums or means calculated on purely metric or topological considerations. Spatial weights can also be calculated by geostatistical methods, minimizing the variance of the values estimated by the model, depending on the variance observed on the values of the variable to be explained depending on the distance of the individuals. The same principle can therefore be applied to regression: local estimation at a point P depends on the value of the explanatory variables at point P and the values of the explanatory variables at neighboring points, with weights calculated by methods of estimation by minimization of the variance in geostatistics.

6.4. An example: analysis of tuberculosis risk factors

This example aims to analyze the importance of socio-economic factors in the incidence of tuberculosis (TB) between 2004 and 2008 in Sisaket Province in Thailand (Figure 6.4), based on epidemiological and demographic data. Sisaket is the province in Thailand with the highest incidence of TB. Epidemiological data were collected in the hospitals.

The incidence rate and SMR per rural and urban sub-districts (216 units in the province) have been calculated from individual patient data. SMR is calculated by age and gender indirect standardization in each district. All the risk analyses then concern these rates by sub-districts and not on the individuals, since epidemiological data are anonymized and they do not contain socio-economic information. Various relationships and sometimes inverse relationships are possible at the level of individuals, and the results of this analysis should not be inferred to individuals. The objective of the study is to provide public health authorities with elements for operational actions, allowing the reduction of the disease incidence in the province, focusing simultaneously on the most affected sub-districts, the most important risk factors and the most vulnerable sub-districts. This example relies on [HAS 15, HAS 18].

6.4.1. *Epidemiological and socio-economic data*

Epidemiological data have been collected with many difficulties from case reports made by public hospitals between 2004 and 2008. Some of the registries were only available in handwritten form and others had been damaged. The reported cases concern all confirmed cases, for all TB types (pulmonary and extra-pulmonary TB). The patients are characterized by their status: new case, relapse, failure and treatment after failure. The cases considered in this study only concern new patients and relapse patients.

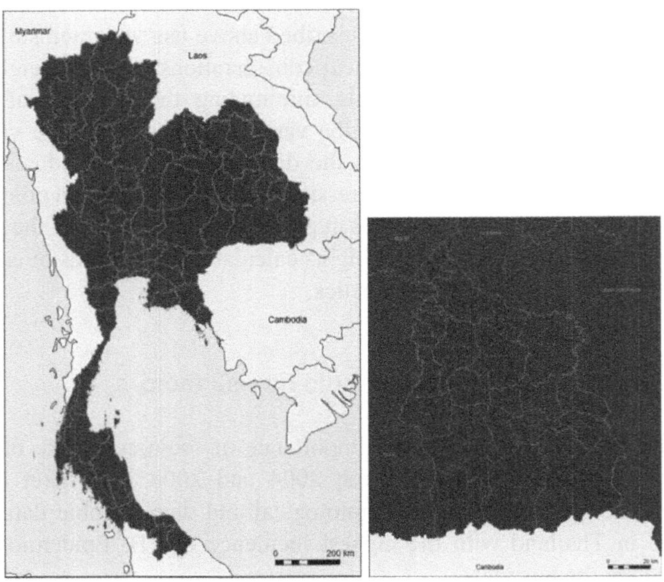

Figure 6.4. *Sisaket Province in Thailand borders Cambodia. For a color version of this figure, see www.iste.co.uk/souris/epidemiology.zip*

Demographic and socio-economic data have been selected from databases at the level of sub-districts and municipalities: Basic Minimum Need, National Rural Development Committee Survey, National Economic and Social Development Board, and National Statistic Office.

The incidence rate calculated from the collections of cases and population data is age and gender-standardized using seven age classes (0–14,15–24, 25–34, 35–44, 45–54, 55–64, > 64) and an indirect standardization with rates calculated over the set of cases reported at the national level. For each sub-district, an SMR and an adjusted incidence are obtained.

Before approaching the study of risk factors, we analyze the statistical and spatial distribution of the incidence observed by sub-district (SMR and adjusted incidence). Indeed, the short-term objective of public health policies should be incidence reduction and treatment improvement, independently of risk factors and their role in explaining the disease, while their long-term objective should be the reduction of vulnerability related to risk factors. It is therefore important to identify the most affected territories, which should be the target of short-term priority public health actions. It is also important to characterize the most vulnerable territories, which should be the target of more long-term actions aimed to reduce this vulnerability. Therefore, this study has two stages: one of them is the analysis of the spatial distribution of incidence, and the other is the search for a model explaining the incidence based on socio-economic risk factors, the spatial analysis of both the model residue and of the most decisive risk factors. If a global model is not satisfactory, a local model will be proposed for the analysis of the spatial distribution of the contribution of each risk factor to the global model.

6.4.2. *Analysis of the statistical and spatial distribution of rates*

QUESTION.– What is the place with the highest number of TB cases? What is the place with the highest SMR of TB? Are these values significant with respect to the set of districts? Does the disease present a particular global spatial distribution, a spatial trend, “hot spots” or clusters?

The analysis of the statistical distribution of SMR at the level of sub-districts shows a rather normal distribution, centered on 1 and varying from 0.2 to 2.7 (Figure 6.5). The distribution nevertheless shows an overdispersion compared to a random distribution (Potthoff–Whittinghill test). The Breslow–Day test highlights the sub-districts whose values of SMR are significantly lower or higher than expected.

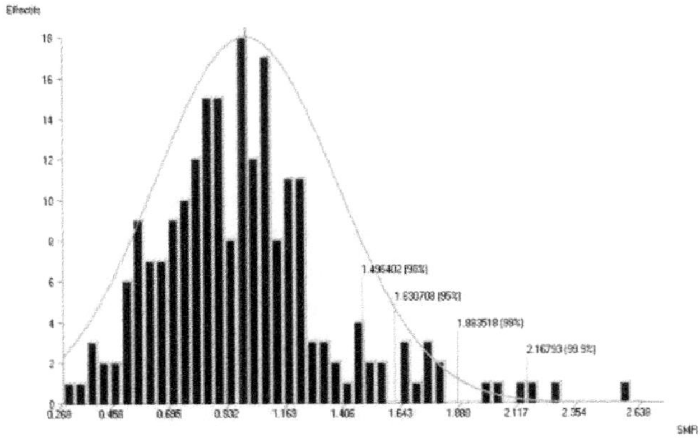

Figure 6.5. *Distribution of incidence and SMR by sub-district in Sisaket Province*

Mapping of adjusted incidence and SMR by sub-district shows a very marked spatial trend, with higher values at the center of the province according to a south-west/north-east axis, with a homogeneous zone of lower values in the west (Figures 6.6 and 6.7). It can also be noted that most of the units whose SMR is significantly lower or higher (Breslow–Day test) are contiguous.

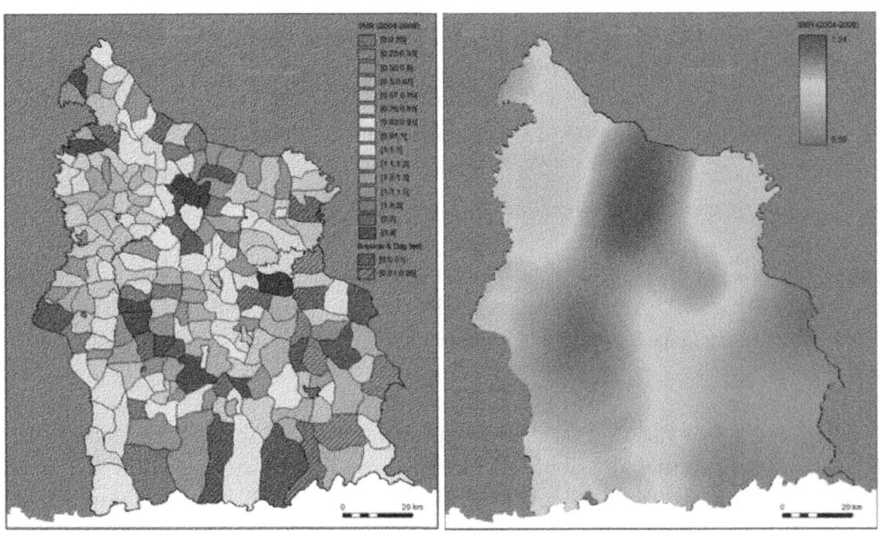

Figure 6.6. *SMR mapping by sub-district, by unit on the left and by trend surface on the right. A clear south-west/north-east trend can be noted, with high values at the center of the province and low values in the east. The hatched zones correspond to statistically significant values of SMR. For a color version of this figure, see www.iste.co.uk/souris/epidemiology.zip*

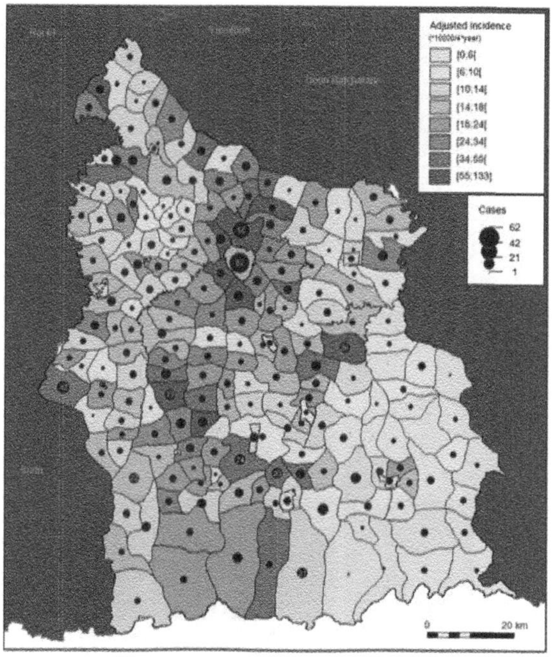

Figure 6.7. *Mapping of the adjusted incidence with the number of cases by sub-district. For a color version of this figure, see www.iste.co.uk/souris/epidemiology.zip*

Spatial analysis of SMR shows a very marked global spatial autocorrelation (highly significant Moran's index, p-value < 0.00001). The Getis–Ord local indicator of spatial association shows the locations of significant points (Z-Score > 1.96 or Z-Score < −1.96) (Figure 6.8). Finally, it is not surprising that the most significant clusters (high values of SMR) calculated by spatial scan (Kulldorff method and Bernoulli model) correspond to the high values occurring on SMR mapping (Figure 6.9).

TB being *a priori* linked to socio-demographic risks, this spatial distribution may be due to the spatial distribution of socio-economic factors and not to contagion relationships (in the broad sense) between geographical units. It is therefore relevant to first try to explain SMR with these socio-economic risk factors, to build a statistical model of SMR depending on these factors, and then once again to do a spatial analysis of the residue in order to verify that the spatial distribution was actually generated by these factors. If there is still a residual autocorrelation, then an autocorrelation term should be added to the model (and therefore an autoregressive SAR model should be used), or one should try to determine whether an environmental factor has been omitted.

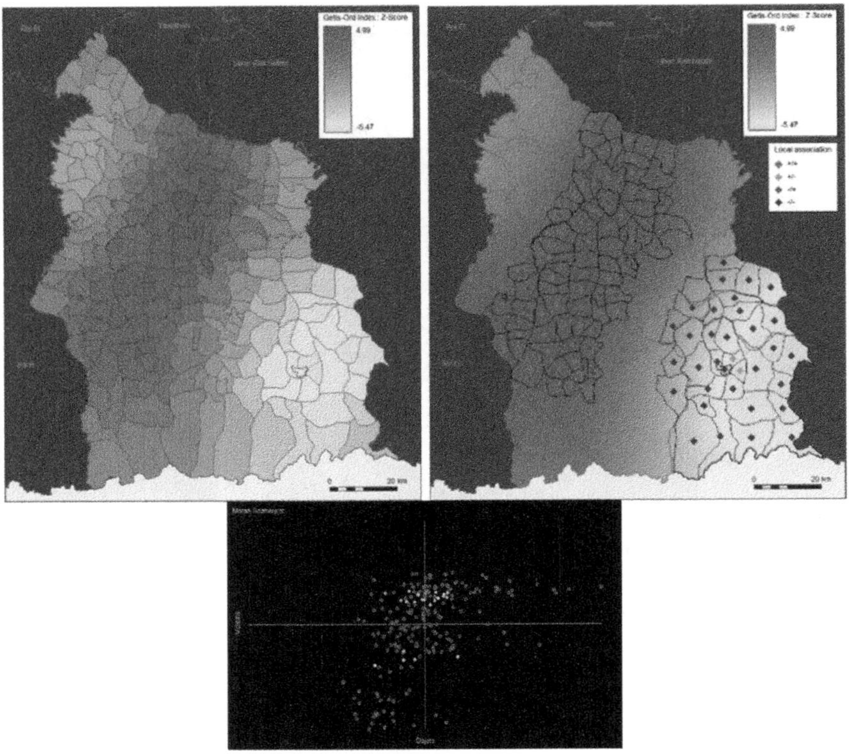

Figure 6.8. *Getis–Ord local indicator of spatial association and the corresponding Moran scatter plot. The points in yellow and red correspond to sub-districts with statistically significant indices. For a color version of this figure, see www.iste.co.uk/souris/epidemiology.zip*

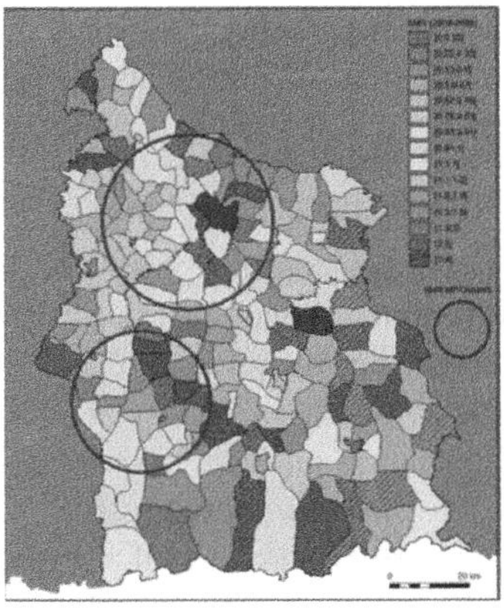

Figure 6.9. *The most significant clusters for high SMR at the level of sub-districts. For a color version of this figure, see www.iste.co.uk/souris/epidemiology.zip*

6.4.3. *Statistical modeling of SMR and incidence*

QUESTION.– Globally, what are the variables that may explain the incidence of TB by sub-district? What is the most accurate statistical model that explains the observed incidence?

The objective here is to determine the parameters of a statistical model explaining the incidence of a disease by sub-district. It is a matter of finding the best statistical model, the most significant risk factors, and measuring their role in the model: what is the best form of the model (multilinear, logistic, logarithmic)? What risk factors (explanatory variables) should be considered? How important is each of them in the model?

The analysis of the residue of the regression also helps verify whether the statistical model allows the explanation of the spatial distribution, particularly the spatial autocorrelation and the low and high incidence zones.

The factors that explain TB are well-known and have been abundantly commented on in the literature (Figure 6.10). They concern several stages of the disease: exposure to the bacterium, risk of developing an infection when exposed to a source of infection, progression of the disease in the body and factors that lead to the aggravation of the disease or death. SMR or adjusted incidence is modeled, as age and gender are known risk factors that cannot be influenced. Furthermore, since this study concerns the socio-economic and environmental risk factors, it is preferable to standardize data rather than to include age and gender in the model. Once the SMR analysis is conducted, it is still possible to revisit the counts or a non-adjusted incidence by considering in the model the structure of the population in each unit in order to reintegrate age and gender for risk assessment purposes.

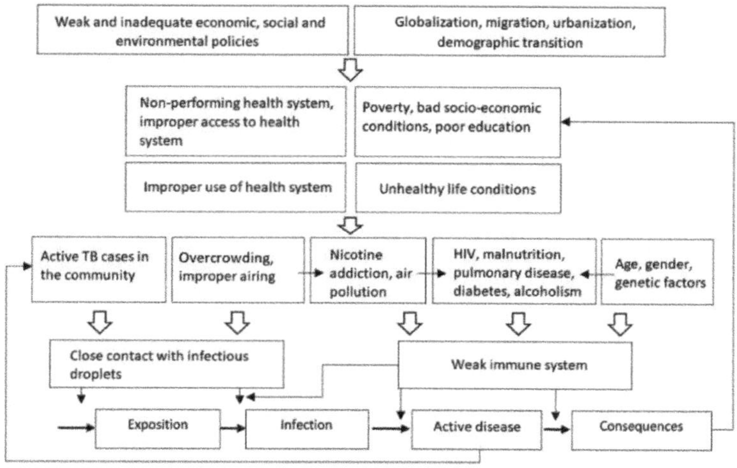

Figure 6.10. *Socio-economic and environmental risk factors of TB (HAS 15)*

The following 12 factors are considered here, expressed in percentage for each district: alcoholism, food quality, dwelling occupancy status, dwelling hygiene, pollution in the dwelling, savings, access to water supply, access to power supply, average income, employment rate for 18–60 year old persons, percentage of household living off rice culture and percentage of households with permanent access to the main road network.

These factors show significant collinearities and they are grouped into five composite factors after exploratory factor analysis with the Kaiser-Meyer-Olkin measure of sampling adequacy: living conditions (variables: alcoholism, nutrition, occupancy status, pollution, hygiene, income), facilities (access to water supply, access to power supply), economic conditions (average income, employment rate), activity (rice culture) and infrastructure (access to road network). These new variables no longer show collinearity, and make it possible to explain approximately 65% of the total variance of the 12 initial variables.

Living Conditions = 0.883(Alcoholism) + 0.658(Nutrition) + 0.724(Hygiene) + 0.483(Pollution) + 0.711(Hygiene) + 0.551(Savings)

Facilities = 0.677(Electricity) + 0.862(Water)

Economy = 0.732(Income) + 0.760(Employment)

Activity = (Rice)

Infrastructure = (Road network)

The mapping of these composite variables does not highlight a spatial distribution allowing the visual explanation of TB incidence distribution.

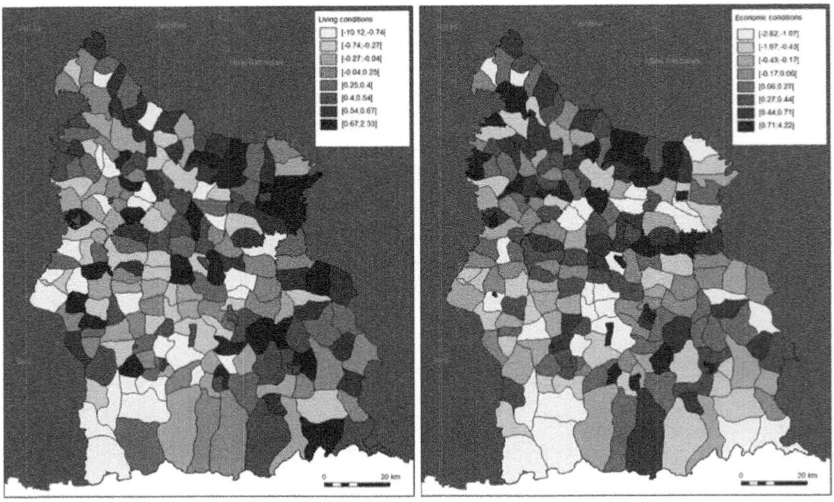

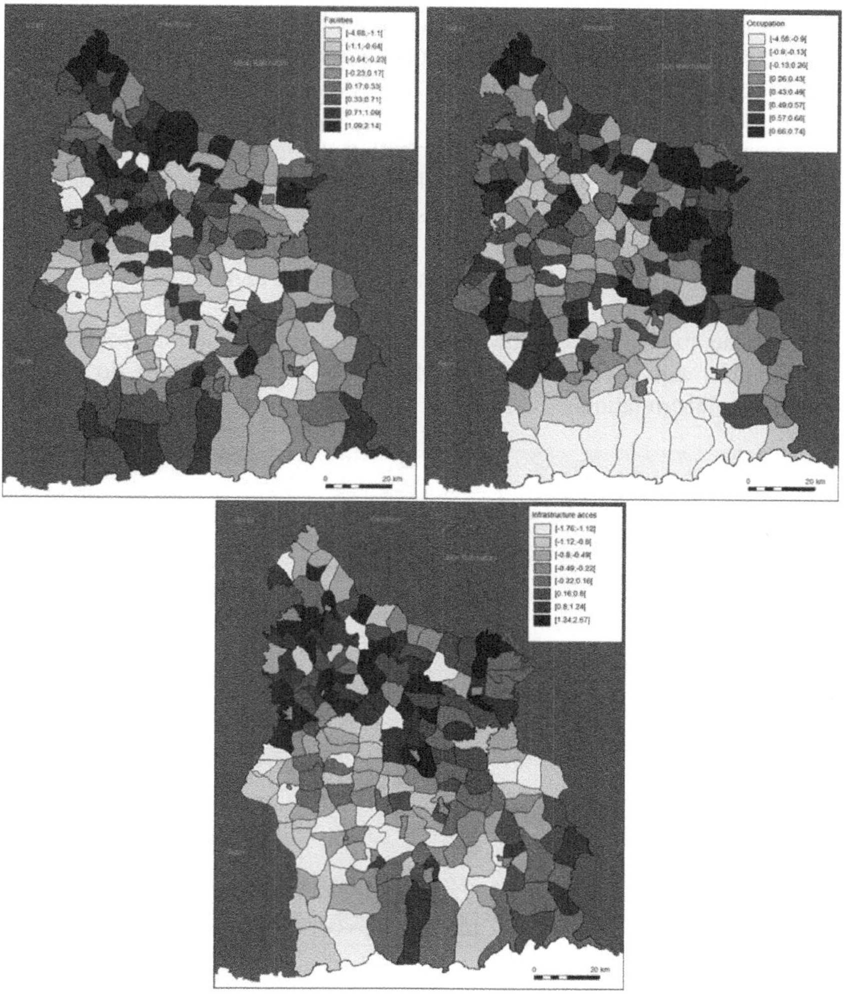

Figure 6.11. *Various composite risk factors considered in statistical modeling (living conditions, economic conditions, facilities, occupation and infrastructure). For a color version of this figure, see www.iste.co.uk/souris/epidemiology.zip*

Based on these risk factors, several models are tested in order to explain SMR and incidence: a multilinear model, a generalized multilinear (logistic) model and a logarithmic (Poisson) model. All the calculations have been made with the SPSS software program.

QUESTION.– What is the quality level of these statistical models?

The quality of the models is assessed by analyzing the model residue: mean, standard deviation and statistical distribution. Various tests are used, depending on the model (chi-square, Wald, R^2, AIC, etc.).

The focus here is on the spatial distribution of the residue, in order to verify whether the spatial autocorrelation noted for SMR or the adjusted incidence has been explained by the model. In the case of the residue, the objective is to see whether the spatial autocorrelation of TB has been absorbed by the regression and therefore only results from risk factors considered in the regressions.

For all the models, it can be noted that the residue still presents a spatial autocorrelation, with a very significant Moran's index. Risk factors do not help to explain this autocorrelation.

Figure 6.12. *The residue of the regression (generalized linear regression, SMR on the left, incidence on the right) presents the spatial autocorrelation. For a color version of this figure, see www.iste.co.uk/souris/epidemiology.zip*

QUESTION.– Can the model be improved by considering spatial dependence?

Since explanatory factors cannot explain spatial dependence, the model can be improved using an autoregressive (SAR) model to introduce an autocorrelation factor (spatial lag). This autocorrelation factor allows the absorption of the spatial dependence but does not explain it. We once again analyze the residue to see whether this SAR autoregressive model improves the quality of the model (statistical characteristics and distribution of the residue). The SAR model is either multilinear, or logistic, or logarithmic (Poisson), with an additional factor of spatial autocorrelation that is included or is outside the link.

The objective is to identify the regression that most reduces this autocorrelation or reduces the range of the residual spatial dependence.

Figure 6.13. *Residue of SAR autoregressive models (linear autoregressive model) and PAR (Poisson autoregressive model). These models no longer present a statistically significant spatial autocorrelation (Moran's index). For a color version of this figure, see www.iste.co.uk/souris/epidemiology.zip*

QUESTION.– What is the distribution of each risk factor detected in the model (irrespective of age and gender)? What are the most affected districts? Are the values spatially significant? Are there any hot spots or clusters of districts for these high values?

The places with the highest TB are not necessarily those where public health action can be the most efficient. Long-term public action involves the reduction of a risk factor, and spatial analysis allows the determination of places where this risk factor should be acted upon as a priority. A further objective is to quantify the influence of this action on the disease, using the model while considering the impact of this action on the value of risk factors.

For this purpose, the spatial characteristics of each risk factor are studied in the order of their importance in the model (mapping, spatial analysis of each risk factor: high-value zones, significance, autocorrelation, hot spots and clusters).

QUESTION.– Are there any spatial variations in the importance of factors, if the regression can vary in space? How does each factor vary? Is there any correspondence between the local importance of a risk factor and the spatial distribution of this risk factor?

The first question corresponds to the relative local importance of a factor compared to another factor, and the second corresponds to the absolute local importance factor, independently of the others. If there is correspondence, this means that public health actions can focus on this factor and on the districts where it is the highest.

The disease is once again modeled (incidence or SMR, depending on whether or not age and gender are introduced in the model) using GWR and GLWR models. The variation of risk factor coefficients in the model is subjected to mapping and spatial analysis, and compared to the spatial distribution of factors themselves. Public health actions should focus on places where risk factors are the highest and where their influence on regression is the strongest, the zones where these two criteria are both met being considered as a priority.

SUMMARY.–

– The operation of spatial aggregation transforms individual data into variables of immobile geographical units, which allows many calculations in these units, and particularly an assessment of probabilities.

– Comparison of geographical units nevertheless poses many problems.

– Standardization of aggregated data is required when a known risk factor must be eliminated.

– Spatial aggregation allows correlation calculations with the environmental variables characterizing the clustering units.

– In contrast to aggregation, the spatial belonging procedure makes it possible to assign environmental or contextual data to individuals and to reinforce the search for a statistical model of risk at the individual level.

7

Space–time Analyses and Modeling

In epidemiology, the availability of time for each event is a significant asset in spatial analyses. Taking time into account provides the possibility of ignoring proximity spatial relationships that should not be taken into account and that might falsify spatial analyses.

Retrospective spatio-temporal analyses essentially concern the following:

– global spatio-temporal analysis (correlation between time and distance, spatio-temporal autocorrelation);

– analysis of the movement of synthesis elements (mobile mean points);

– search for clusters and local spatio-temporal hot spots and their evolution in time;

– study of emergence and diffusion (finding the cases that generate diffusion, which are called index cases, studying the geometric parameters of diffusion);

– pathway reconstruction, for the reconstruction of epidemics propagation;

– spatio-temporal modeling, particularly modeling of epidemics.

Time series analyses are commonly used in many fields (especially in meteorology, natural or agricultural resources, oceanography, economics, etc.). In epidemiology, the retrospective spatio-temporal analysis always starts with a global temporal study: global trend, search for seasonal variations, cycle, etc.

7.1. Time–distance relationships

QUESTION.– Globally, are two events that are close in time also close in space?

This is the first question, the preliminary stage allowing the consideration of a contagion relationship between health events. It involves an analysis of the spatial autocorrelation of time. Many tests have been developed for this purpose (Knox, Mantel, space-time K-functions, etc.).

For example, after having fixed a duration and a distance, the Knox test uses all the pairs of events to establish a table of contingency between the difference in time and the difference in space, then applies a χ^2 test to see whether the values found are significantly different from a random distribution [KNO 64] (Figure 7.1).

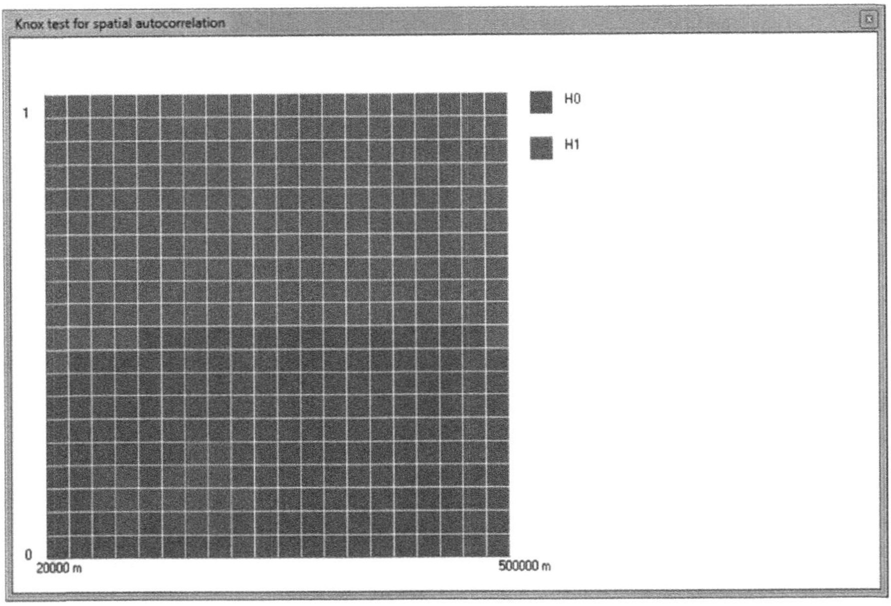

Figure 7.1. *Graph indicating which are the values of differences in distance (on the x-axis) and in time (on the y-axis) for which the Knox test allows the rejection of H_0*

SavGIS	The Knox test is available with the menu: Stat → Spatial statistics → Knox extent

Table 7.1. *In practice*

7.2. Mobile mean points

QUESTION.– Does the global movement of the phenomenon have a specific form? Is it centered on a place or does it move in time?

The idea is simple: rather than considering the set of events, a mean (or median) center is considered for the events belonging to a temporal window with fixed duration. The movement of this temporal window in time results in a series of mean centers, which allows the smoothing and synthesis of the global movement of the phenomenon (Figure 7.2).

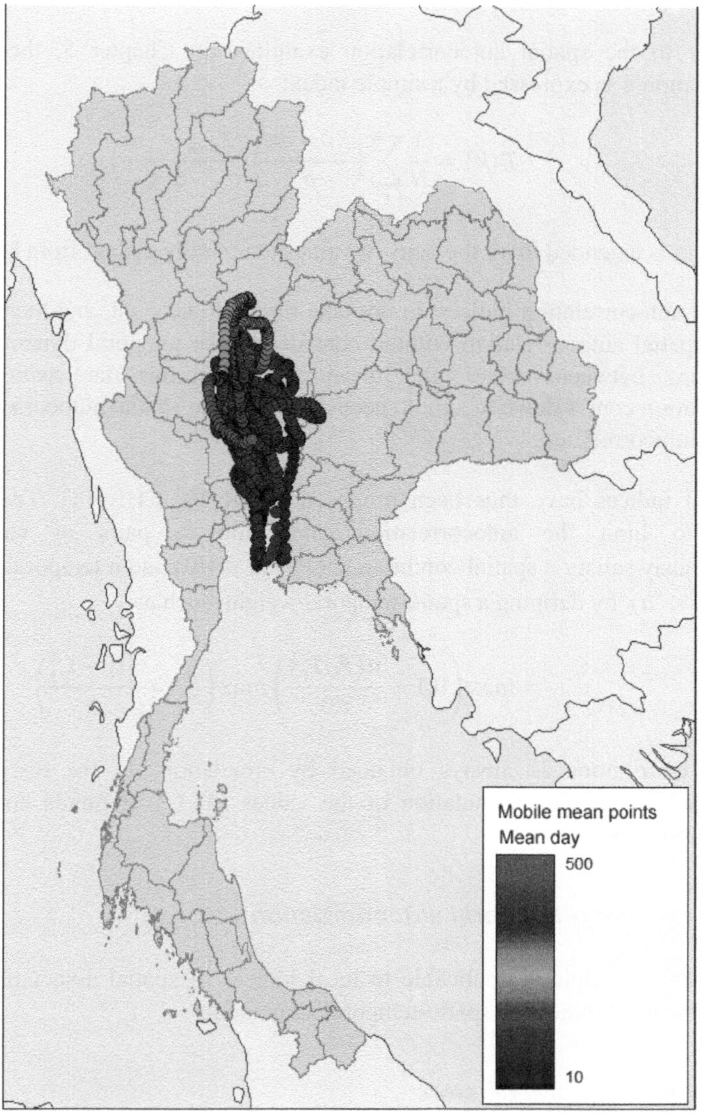

Figure 7.2. *Movement of mean centers in time during a wave of avian influenza in Thailand*

7.3. Spatio-temporal autocorrelation and clusters

7.3.1. *Global spatio-temporal autocorrelation*

QUESTION.– Globally, are the values of objects correlated with the proximity of objects in time? What about in time and space?

Similar to the spatial autocorrelation examined in Chapter 5, the temporal autocorrelation T is expressed by a simple index:

$$T(k) = \frac{1}{N}\sum_{t}\left(\frac{X_t - m}{\sigma}\right)\left(\frac{X_{t-k} - m}{\sigma}\right)$$

The sum is extended to all the pairs of values temporally distant from k.

Spatial autocorrelation indices do not take time into account, and therefore they measure spatial autocorrelation without considering the temporal dynamics of the relationships between values and distances. Measuring the spatio-temporal autocorrelation comes down to simultaneously measuring spatial autocorrelation and temporal autocorrelation.

Several indices have thus been proposed [HAR 10, CHE 11]. The simplest idea is to limit the autocorrelation calculation to pairs of values that simultaneously satisfy a spatial condition ($d(P_i, P_j) < d0$) and a temporal condition ($|t_1 - t_2| < h$), by defining a spatio-temporal weight, such as:

$$w_{i,j} = \max\left(0.1 - \frac{d(P_i, P_j)}{d0}\right)\max\left(0.1 - \frac{|t_i - t_j|}{h}\right)$$

Index distribution is always obtained by simulation. In the discrete case, simulation involves the permutation of the values (X_i, t_i) of points among their spatio-temporal positions.

7.3.2. *Local spatio-temporal autocorrelation*

The same principle is applicable to local indices of spatial association, which then become local indices of spatio-temporal association.

7.3.3. *Spatio-temporal clusters*

QUESTION.– Are there clusters of objects/values that appear or disappear in time? How do these clusters move?

Two types of studies use the search for spatio-temporal clusters: on the one hand, retrospective studies, where all data are available, and on the other hand, surveillance systems, where data are collected as the study progresses, and warnings are triggered when an "abnormal" situation is detected [ROG 97].

The search for spatio-temporal clusters relies on the same principle as for mobile mean points: in the search for the cluster, only the events of a temporal window are considered. The temporal window moves in time, which makes it possible to detect the appearance or disappearance of clusters. Spatial scan methods are the most commonly used [KUL 97].

Warning systems for infectious diseases use these methods; warning and surveillance systems often involve a real-time search for the occurrence of a spatial cluster.

It is at the beginning of an epidemic, when there are few events, and they are grouped due to their proximity, that clusters are statistically the most significant and their detection has the strongest impact on epidemic reduction.

The SaTScan™ software program is specialized in the search for clusters by spatial scan.

SavGIS	The calculation of the spatio-temporal cluster is available with the menu: Calculate → Geostatistical calculation → Spatio-temporal clusters
SaTScan	The SaTScan™ software program is specialized in the search for clusters by spatial scan

Table 7.2. *In practice*

7.3.4. *Statistical modeling: GTWR*

Spatio-temporal extensions of the spatially weighted regression method GWR have been developed to include time in the analysis of non-stationary processes. The spatially weighted regression methods presented in Chapter 6 rely on the hypothesis of constant relationships in time. In order to take time into account, several authors have proposed to build spatio-temporal weight matrices for SAR or GWR models [WU 14, YU 14] using spatio-temporal distances to measure the distances between observations. Other authors have proposed the division of time into mobile temporal windows in order to determine the coefficients of GWR models using only the data observed in these temporal windows [FOT 15].

7.4. Emergence, diffusion, pathway

QUESTION.– Is it possible to separate the emergence phenomena linked to relationships with the environment from the contagion phenomena linked to relationships of proximity between cases?

QUESTION.– Is it possible to reconstruct and characterize the pathway of a contagion?

As already highlighted, when searching for the risk factors of a disease based on retrospective data, it is important to separate factors that pertain to a spatial relationship with the environment (mobile or immobile) from factors resulting from spatial relationships between agents. This is particularly important when the spatial relationships between agents are significant in the phenomenon, which is the case of infectious diseases. The analysis of the relationship between disease occurrence and environment by taking into account the set of cases involves the consideration of the contagion process in the simulation of random situations under null hypothesis, which is often difficult.

A further solution involves ignoring the cases due to relationships between agents (contagion). The idea is therefore to separate what corresponds to disease emergence (and therefore *a priori* due to an environmental factor) from what corresponds to diffusion by contagion, and to limit the analysis of health–environment relationships only to events that correspond to the emergence. On the contrary, diffusion by contagion can itself be influenced by the environment (by all the density-related factors: transportation, schools, markets, etc.). To study it, it is important to ignore the cases due to contagion.

Data that lead to establishing a direct causal relationship between one case and the previous case are rarely available. When the time and location of cases are available, these relationships can be reconstructed using a spatio-temporal calculation. The objective is double:

– separate emergence and diffusion by proximity;

– reconstruct the pathway of the epidemic and deduce its space–time characteristics from the pathway.

Pathway reconstruction or the characterization of emergence must be considered as modeling (Figures 7.3 and 7.4). Complementary data can also be used in the calculation, such as genetic filiations of pathogens, in order to refine the model and not limit it to purely geometric and temporal rules.

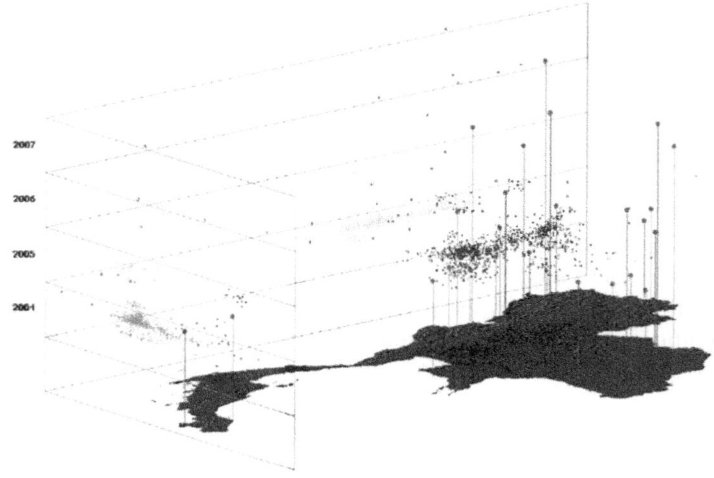

Figure 7.3. *Avian influenza cases in Thailand, three-dimensional image. Time is represented on the z-axis. Red points correspond to the points of emergence (no previous cases for one month within a 50 km radius)*

SavGIS	Emergence and diffusion calculations are available with the menu: Calculate → Spatio-temporal calculations

Table 7.3. *In practice*

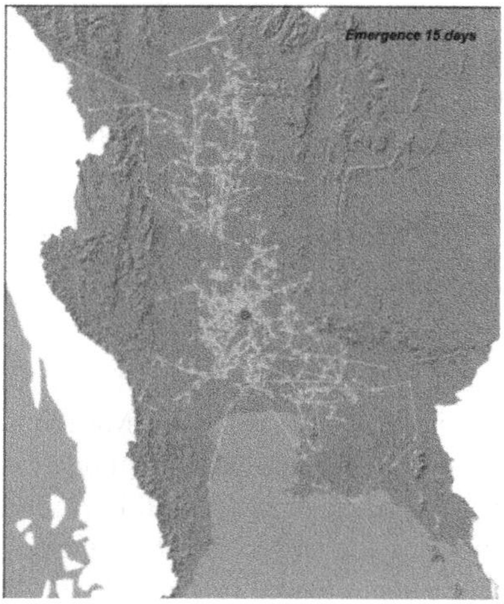

Figure 7.4. *Pathway reconstruction for avian influenza epidemic in Thailand, from a point of emergence (index case). Reconstruction is made by the backward chaining depending on the distance between cases and an assumed propagation speed*

7.5. Spatio-temporal modeling of health phenomena

7.5.1. *Process modeling and simulation*

Spatio-temporal modeling can be considered as the final stage in the analysis of a health phenomenon. Spatio-temporal modeling allows the approximation of the probability distributions of the phenomenon parameters, taking into account the model intrinsic characteristics and, by determining the spatial processes unfolding in the phenomenon, the simulation of its evolution in time. Probability distributions are obtained mathematically or by using simulations from the theoretical model and the initial conditions. In epidemiology, modeling has been used for several centuries (Figure 7.5).

A model is referred to as spatial if location (of agents or populations) is among the input variables, and as spatio-temporal if time is also an input variable. The model can only yield a global, non-localized result (such as global incidence, risk, transmission rate, etc.), a temporal result on phenomenon dynamics (for example, the global evolution in time of an epidemic), a localized result when location is an output parameter of the model (for example, an incidence in a geographical unit), or still spatio-temporal when the output dynamics is itself localized (for example, incidence evolution in a geographical unit in time).

From the modeler's perspective, spatial analysis of the observed phenomena first aims to determine some characteristics and parameters of the searched model, and particularly the characteristics of the processes related to relationships and interactions among agents and between agents and their environment. Spatial analysis of the observed situations can also have a role in validating the model, comparing the spatial characteristics of the modeled situations with the spatial characteristics of the observed situation. Finally, spatial analysis of the results of a spatial model obtained from differential equations and diffusion equations may help in the characterization of underlying theoretical global spatial processes. All of the analysis methods presented in Chapters 5 and 6 can be applied to the result of a model.

The main theoretical models in the generation of events in spatial processes have already been mentioned: Poisson, Neyman–Scott, Matern, Cox, Gibbs, etc. Most of these purely spatial processes have been built for modeling continuous physical or environmental phenomena in a dense space (natural resources, meteorology, botany, etc.). In epidemiology, most of the observed phenomena are limited to a discrete support (places where the phenomenon is possible or those in which they are aggregated), and spatial analysis (of observed situations as modeled situations) provides indications on and characteristics of the underlying global theoretical processes, independently of the spatial distribution of the support.

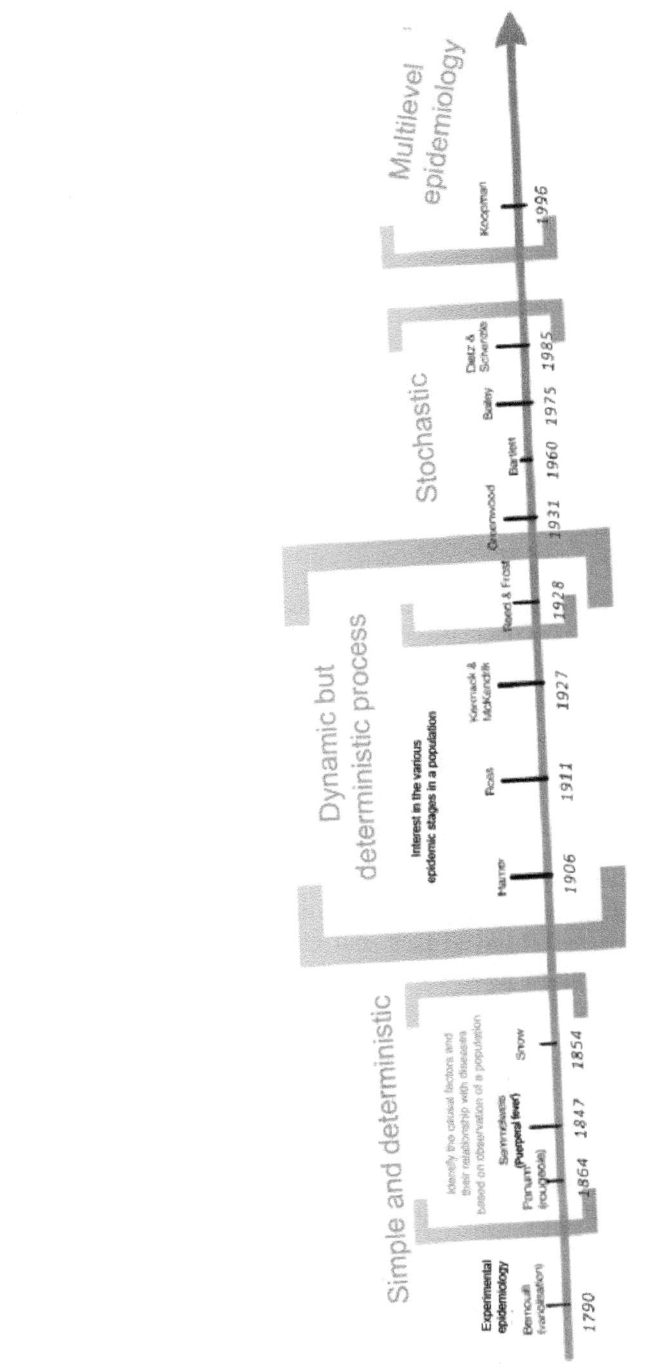

Figure 7.5. *The evolution of modeling methods in epidemiology (according to [MAN 16])*

Spatio-temporal modeling differs from statistical modeling. The objective of statistical modeling is a mathematical formulation of a parameter or risk, depending on assumed risk factors, and the coefficients of which are determined by statistical or heuristic analysis that only relies on the observed data. Because it approaches the phenomenon as a whole, spatio-temporal modeling is *a priori* much richer. However, it should be realistic, and therefore it should take into account as many parameters as possible, considering a health phenomenon as a complex system, as described in Chapter 1. These parameters include spatial relationships and interactions between agents (hosts, vectors, reservoirs, pathogens) and between the agents and their environments.

When studying infectious diseases, spatio-temporal modeling is particularly useful in understanding and anticipating epidemics. Generally, it involves a characterization of individuals in various states (SEIR "compartmental" model: susceptible, exposed, infectious, recovered or deceased) and efforts are made to model major phenomena that support the passage of an individual from one state to another. When a model takes into account numerous processes, it can rapidly become very complex.

Modeling first of all implies knowledge and understanding of the operation of agents involved in the system. The large majority of models are assumed simplifications of reality. Mathematical solutions or simulations allow the determination of the distributions of probability of variables, the dynamics of a variable in time and sometimes the localization of the result in time and space, when the result of the model is itself spatialized [DIE 00, AUC 08].

Two large categories of methods are used in modeling:

– A deterministic approach, which relies on differential equations (modeling the inputs and outputs of a state), and more generally an analytical approach of which the parameters result from the observed data and characteristics of the agents. Models that deal with populations instead of individuals have difficulty in taking into account complexity and particularly spatial relationships.

– A non-deterministic approach, which directly models the evolution of agents, whose individual behavior is described by the rules that experts formulate based on systemic analysis (individual-based and multi-agent models). The state of each agent is calculated at each time step from their behavior, environment and relationships between the agent and the set of other agents. These models provide the possibility of a much more realistic description of the studied phenomenon, close to the complex system that finely describes reality. They make it possible to consider spatial relationships at each time step. These models involve heavy calculations and their use has been facilitated by the development of computational capacities. The main difficulties are rule determination and validation.

7.5.2. *The deterministic approach of SEIR models*

The SEIR model is specific to epidemiology. It characterizes agents by four different states: susceptible, exposed, infected and recovered. At each time step, susceptible subjects may enter the system, and recovered and deceased subjects may exit the system. A "compartment" is defined as the population of agents that are in the same state. The model relates to the dynamics in a population of the count of the agents, for each state; the result concerns only the evolution of the count of each compartment in time. Modeling involves the quantification of the number of passages from one state to another by unit of time, depending on the processes that describe the phenomenon.

All agents of the same type (host, vector, reservoir, etc.) that are in the same state are considered as identical and undergo the same processes in the passage from one state to another; a compartment is considered as homogeneous. Processes concern individuals, but the parameters describing these processes are evaluated by statistics on the phenomena observed on the populations, and therefore result from means or variances. Agent location is not involved in the modeling, which considers the population as a whole; individuals are subjected to the same environmental conditions, wherever they are, and interactions are expressed by processes that take into account the location of agents and of their interactions only by means or random processes. The model is expressed by a mathematical formulation, and the result is only determined by parameters and initial conditions. Random components can be introduced in the coefficients of some parameters by studying the variability in the observed data.

For example, the highly simplified SEIR modeling of malaria published by Ross in 1910 involves two types of agents: human individuals and (anopheles) mosquitoes, which can bite human individuals and transmit the malaria parasite or can become infected themselves if they bite an infectious human individual. Similar to mosquitoes, human individuals are grouped into two compartments: susceptible and infectious. Compartments are presumed uniform (all individuals have the same risk of being bitten by a mosquito and mosquitoes have the same risk of being infected). For each time step, the transfers from the compartment of susceptible individuals to the compartment of infectious individuals and vice versa are evaluated. An infected individual is assumed recovered after some time and joins the compartment of susceptible individuals. An infected mosquito is assumed to remain infected throughout its life, and all mosquitoes have the same life expectancy. This highly simplified model has nevertheless contributed to advancing two concepts that are essential for the strategies to fight against vector-borne diseases: the basic reproduction number R0, which corresponds to the mean number of individuals that an infected individual may infect, and which can be used to determine the evolution of an epidemic in a population; and the vector capacity, which summarizes all the characteristics of the vectors impacting the model and R0 [REI 13].

Human individuals: a mosquito is assumed to bite a human individuals per unit of time Δt. Therefore, a number of $I_v(t)$ infectious mosquitoes will make $aI_v(t)$ bites per unit of time Δt.

The probability of an individual becoming infectious after an infecting bite is assumed to be b_1.

There are $S_h(t)$ susceptible human individuals, and the number of infectious human individuals is $I_h(t) = H - S_h(t)$, where H is the global population (assumed constant in time). The number of new infectious individuals in Δt is therefore equal to: $b_1 a I_v(t) \frac{I_h(t)}{H} \Delta t$. An individual's mean recovery speed is assumed to be γ_h per unit of time, and mortality is assumed μ_h. The following differential equation is readily obtained:

$$\frac{dI_h(t)}{dt} = b_1 a I_v(t) \frac{H - I_h(t)}{H} - (\gamma_h + \mu_h) I_h(t)$$

For mosquitoes: a mosquito that bites an infectious human individual has a probability of getting infected of b_2. The mosquito population is denoted by V (assumed constant in time). This leads to:

$$\frac{dI_v(t)}{dt} = b_2 a (V - I_v(t)) \frac{I_h(t)}{H} - (\gamma_v + \mu_v) I_v(t)$$

Considering $x(t) = \frac{I_h(t)}{H}, y(t) = \frac{I_v(t)}{V}$ (prevalences at time t), and $\dot{x(t)} = \frac{dx(t)}{dt}$ and $\dot{y(t)} = \frac{dy(t)}{V}$ (instantaneous incidences), considering $m = {}^{V}/_{H}$ and neglecting μ_h and γ_v, the Ross equations are obtained:

$$\dot{x(t)} = mab_1(1 - x(t))y(t) - \gamma_h x(t)$$

$$\dot{y(t)} = ab_2(1 - y(t))x(t) - \mu_v y(t)$$

Ross equations are a particular case of Lotka–Volterra equations used for modeling prey–predator relationships.

Special attention is paid to the system's equilibrium points, with null instantaneous incidence $\dot{x} = \dot{y} = 0$. A clear equilibrium point is $x = y = 0$: it is the no-disease equilibrium. When we search for an equilibrium point with $x \neq 0$ and $y \neq 0$, the following expressions are obtained:

$$x = \frac{\frac{ma^2 b_1 b_2}{\mu\gamma} - 1}{\frac{ma^2 b_1 b_2}{\mu\gamma} + \frac{b_2 a}{\mu}} \text{ and } y = \frac{\frac{ma^2 b_1 b_2}{\mu\gamma} - 1}{\frac{ma^2 b_1 b_2}{\mu\gamma} + \frac{mb_1 a}{\gamma}}$$

The quantity $\frac{ma^2 b_1 b_2}{\mu\,\gamma}$ corresponds to the basic reproduction number $R0$ and is used to determine the existence of another equilibrium point that differs from 0, called the endemic equilibrium. Model stability around its equilibrium points is studied by the

Lyapunov methods: when $R0 < 1$, the system tends towards equilibrium 0,0, no-disease equilibrium, and this equilibrium point is stable. When $R0 > 1$, the no-disease equilibrium point is not stable and the system tends towards the endemic equilibrium point and the disease is persistent.

Box 7.1. *Ross model*

The initial Ross model has omitted many parameters (for example, latency times between infected and infectious state for both human individual and mosquito, duration of immunity, differences between symptomatic and asymptomatic subjects, etc.). In order to take these parameters into account, the model has been subjected to many developments [MKE 04]. Nevertheless, most compartmental models elaborated during the last 40 years are very similar to the Ross model, despite increasing awareness of the geographical, ecological and epidemiological complexities, which require the differentiation of socio-economic characteristics of the agents and the consideration of the environmental conditions that they are subjected to. Taking this complexity into account requires these models to be spatialized or individualized.

7.5.3. *SEIR models and localization*

The main critique that may be formulated of global SEIR models is specifically that they are global and they only consider means over a population, while the processes at play depend on individual variables. This once again highlights the ecological fallacy problem: for example, in the same population, the susceptible individuals are not necessarily those exposed, and approaching individual mechanisms by using mean values in a population may lead to errors; similarly, an R0 results from a global approach in a population and is a transmission mean per individual, but the reaction in terms of public health is very different, depending on whether the variance is low and the distribution is normal or there are several *super-spreaders*, or more generally a sub-population whose probability of disease transmission is far above that of the global population. This is indeed the case with many infectious diseases.

If the model is required to take into account the differences in the environmental, socio-economic or behavioral conditions that influence processes at the level of individuals, the individuals in the population should be differentiated depending on these conditions and more homogeneous sub-populations should be defined for these variables. Moreover, if the objective is to study the spatial distribution of the modeling result, the model should be "spatialized" and have as output a localization or a spatial value, and consequently sub-populations that are homogeneous in space, corresponding to aggregation into geographical units.

Fortunately, the variables related to the vulnerability of populations often have non-random spatial distributions. The same is true for many environmental risk factors. The solution is to define sub-populations of agents corresponding to a space division that reduces the intra-variance of the main hazard parameters and of vulnerability that is considered as immobile (population susceptibility and immunity, socio-economic conditions, environmental conditions), according to their importance in the model. As for mobile factors, they should be evaluated by a geographical unit of this division at each time step (for example, climate variables or contacts between agents).

Therefore, a division into geographical units should be found, which allows the definition of sub-populations that are expected to be more homogeneous with respect to environmental and socio-economic conditions, and which allows a different parameterization of the processes modeled in each sub-population by a classical SEIR compartmental approach. This is the usual problem of disaggregating a population into groups, with the minimization of inter- and intra-variances between groups.

To be realistic, spatialized SEIR modeling must also take into account the count transfers between the sub-populations in each compartment and therefore between various geographical units. Reaction–diffusion models are then used to take into account the propagation effect and the impact of population movements on disease persistence or extinction. The characteristics of these reaction–diffusion models are evaluated by spatio-temporal analysis of the observed situations. This brings us once again to the concept of metapopulation, of special use in ecology, with models for the management of species or for the modeling of prey–predator systems.

For example, most models of diffusion of an infectious disease must take into account proximity-based diffusion (adjacent units exchange agents in each compartment) as well as long-distance diffusion (representing long-distance jumps and new local occurrences). The rates of contacts between individuals, geographical connectivity between units and the spatial structures of the movements are essential parameters of diffusion models.

7.5.4. *Non-deterministic approach of multi-agent models*

Ultimately, it is possible to have complete disaggregation and the model could consider sub-populations with only one individual: these models are called individual-based models (*IBM*). When several types of agent are involved, these models are also known as multi-agent models (SMA). They have the advantage of considering the actual values of the individuals, avoiding any ecological fallacy and being able to directly address interactions and spatial relationships between agents. The main difficulty is, of course, in knowing the individual values of all agents' parameters, which is generally impossible. In the vast majority of cases, the actual values are assessed by population survey, and then randomly assigned to agents depending on these probability distributions. Moreover, some elements related to individuals are difficult if not impossible to obtain, and are equally difficult to address in a survey, as mentioned in Chapter 1. This is the case of individual movements and also of some events with significant spatial variability.

Finally, these individual-based models can only be solved by a computer simulation; instead of equations built using a mathematical formulation of the process, individual-based modeling uses rules that apply to individuals, at each time step, to modify their attributes. The result of modeling is the statistical description of some variables, and mainly the count dynamics of each compartment in time. These models are non-deterministic to the extent that they cannot be solved analytically. Moreover, most rules involve random elements.

The spatialization of multi-agent models is rather simple, as it is sufficient to take into account the localization and have rules for the movements of individuals. On the contrary, it is very demanding in terms of the quantity of geographical data and especially of calculations. An evaluation of the environmental conditions to which each agent is subjected depending on its location (the corresponding rules being applied to it) is required at each time step. Spatial interactions between agents must also be evaluated (the corresponding rules being applied to them). Most systems use a simplified geographical reality in order to reduce time calculations (for example, solving these geometric calculations in raster form, the environment being spatially described as a grid).

NetLogo is a simplified environment for the development of multi-agent modeling (ccl.northwestern.edu/netlogo/). NetLogo is a free, open-source and user-friendly system. It allows a simplified consideration of the localization of agents using a grid on which agents can move.

GAMA is a free modeling and agent-based simulation platform developed since 2007 (http://gama-plaform.org). It is especially dedicated to the simulation of spatial phenomena.

GAMA integrates a development environment with its agent-based modeling language (GAML), a complete simulation environment including the management of complex visualizations, and offers advanced functionalities:

– advanced management of geographical data that, by means of dedicated spatial operators, provides the possibility of reifying, manipulating and developing geographical data;

– a set of structures and commands that facilitate the definition of multilevel models;

– automated tools that support the exploration of models allowing the definition of experimental plans and their application to high-performance calculation resources (cluster, grid);

– bridges and possibilities of coupling with other tools used in the modeling of complex systems, notably R and OpenMole.

Box 7.2. *Software programs and languages for multi-agent modeling*

Glossary

Choropleth map: map that associates a numerical value of each geographic unit with a color. Numerical values are divided into classes, each geographic unit being painted with a color corresponding to its class.

Cluster: statistically significant spatial grouping of localized objects.

Description scale: geographic-resolution used for schematic representation and description of real-world objects.

Ecological error: an individual is assigned the result obtained for a group to which it belongs.

Epidemiology: the description and measurement of the characteristics of a pathology or of a population's state of health. Study of the factors influencing a population's health and diseases.

Geographic information systems: information system designed to enter, store, manage, analyze and represent all types of localized geographic data.

Geomatics: computer-assisted tools and methods allowing the acquisition, representation and analysis of geographical data.

Health geography: geographic analysis of the health system and of the spatial distribution of diseases.

Host: in epidemiology, the host is the (human or animal) species that is susceptible to infection with a given pathogen.

Monte Carlo simulation: calculation method used for determining the statistical distribution of a numerical value in the case of a null hypothesis. The simulation uses draws following the probability distribution corresponding to the null hypothesis, the set of results allowing the approximation of the sought-after statistical distribution.

Null hypothesis: in statistics, this applies to a hypothesis that is postulated and then submitted to statistical test in an attempt to refute it. This involves the risk of incorrectly refuting it (type I error or alpha risk), or not refuting it when it is false (type II error or beta risk).

Odds ratio: ratio of the odds of the event in the studied (exposed) group to the odds of the event in the unexposed (control) group. For each group, the odds of the event are estimated from the number of cases divided by the number of non-cases.

Pathogen: infectious agent (virus, bacterium, parasite, fungus, prion, etc.) that can cause a health disorder in certain human or animal individuals.

Reservoir: a reservoir is any species that carries and transmits a pathogen, while not necessarily ill themselves (for example, civets in the case of coronaviruses).

Spatial analysis: data analysis which aims to highlight structures, patterns or geometric characteristics of a set of objects, using their localization. Furthermore, spatial analysis focuses on the interactions and processes that create these structures.

Vector: living organism (mosquito, tick, bat, rat, cat, dog, fox, snail, etc.) that allows the transmission of a pathogen to human or animal individuals. The vector itself can be affected by the pathogen. A vector-borne disease is a disease transmitted by a vector.

References

[ABR 13] ABRIAL D., CHARRAS-GARRIDO M., DE GOER DE HERVE J. *et al.*, "Méthode de classification pour la cartographie du risque basée sur les champs de Markov cachés discrets", *Journées d'Animation Scientifique du Département Santé Animale*, Cap d'Agde, available at: http://prodinra.inra.fr/record/257177, 2013.

[ALB 00] ALBERT D., GESLER W., LEVERGOOD B., *Spatial Analysis, GIS, and Remote Sensing Applications in the Health Sciences*, Ann Arbor Press, 2000.

[ANC 02] ANCELLE T., *Epidémiologie et Statistique*, Editions Maloine, 2002.

[ANS 88] ANSELIN L., BERA A.K., "Spatial dependence in spatial regression model, with an introduction to spatial econometrics", in ULLAH A., GILES D.E. (eds), *Handbook of Applied Economic Statistics*, pp. 237–289, Marcel Decker, New York, 1988.

[ANS 95] ANSELIN L., "Local indicators of spatial association – LISA", *Geographical Analysis*, vol. 27, pp. 93–115, 1995.

[ANS 96] ANSELIN L., "The Moran Scatterplot as an ESDA Tool to Assess Local Instability in Spatial Association", in FISCHER M., SCHOLTEN H., UNWIN D. (eds), *Spatial Analytical Perspectives on GIS*, Taylor and Francis, London, pp. 111–125, 1996.

[AUC 08] AUCHINCLOSS A.H., DIEZ-ROUX A., "A new tool for epidemiology: the usefulness of dynamic-agent models in understanding place effects on health", *American Journal of Epidemiology*, vol. 168, no. 1, 2008.

[BAC 63] BACHI R., "Standard distance measure and related methods for spatial analysis", *Papers in Regional Sciences*, vol. 10, no. 1, pp. 73–132, 1963.

[BAR 75] BARTLETT M.S., *The Statistical Analysis of Spatial Pattern*, Chapman and Hall, Andover, 1975.

[BEC 12] BECERRA S., "Vulnérabilité, risques et environnement : l'itinéraire chaotique d'un paradigme sociologique contemporain", *VertigO*, vol. 12, no. 1, available at: http://vertigo.revues.org/11988; doi: 10.4000/vertigo.11988, 2012.

[BEG 00] BEGUIN M., PUMAIN D., *La représentation des données géographiques – Statistiques et cartographie*, Armand Colin, Paris, 2000.

[BER 67] BERTIN J., *Sémiologie graphique*, Mouton/Gauthiers-Villars, Paris, 1967.

[BES 91] BESAG J., NEWELL J., "The detection of clusters in rare diseases", *Journal of the Royal Statistical Society, Series A*, vol. 154, pp. 143–155, 1991.

[BIV 08] BIVAND R., PEBESMA E.J., GÓMEZ-RUBIO V., *Applied Spatial Data Analysis Using R*, Springer, Heidelberg, 2008.

[BIV 17] BIVAND R., ALTMAN M., ANSELIN L. *et al.*, Package 'spdep', available at: https://cran.r-project.org/web/packages/spdep/spdep.pdf, 2017.

[BOR 02] BORDIN P., *SIG, concepts, outils et données*, Hermès, 2002.

[BOU 93] BOUYER J., HÉMON D., CORDIER S. *et al.*, *Epidémiologie, principes et méthodes quantitatives*, Inserm, 1993.

[BOW 97] BOWMAN A.W., AZZALINI A., *Applied Smoothing Techniques for Data Analysis*, Oxford University Press, London, 1997.

[BRE 87] BRESLOW N.E., DAY N.E., *Statistical Methods in Cancer Research, vol. I & II*, WHO, International Agency for Research on Cancer, 1987.

[BRU 87] BRUNET R., *La carte, mode d'emploi*, Fayard/Reclus, Paris, 1987.

[BRU 98] BRUNSDON C., FOTHERINGHAM S., CHARLTON M., "Geographically weighted regression", *Journal of the Royal Statistical Society: Series D (The Statistician)*, vol. 47, pp. 431–443, 1998.

[CHE 11] CHENG T., HAWORTH J., WANG J., "Spatio-temporal autocorrelation of road network data", *Journal of Geographical Systems*, vol. 14, no. 4, pp. 1–25, 2011.

[CLA 87] CLAYTON D., KALDOR J., "Empirical Bayes estimates of age-standardized relative risks for use in disease mapping", *Biometrics*, vol. 43, no. 3, pp. 671–681, 1987.

[CLI 69] CLIFF A.D., ORD J.K., "The problem of spatial autocorrelation", in SCOTT A.J. (ed.), *Studies in Regional Science*, Pion, London, pp. 25–55, 1969.

[CLI 81] CLIFF A.D., ORD J.K., *Spatial Processes: Models and Applications*, Pion Limited, 1981.

[CRE 93] CRESSIE N.A., *Statistics for Spatial Data*, John Wiley & Sons, Inc., 1993.

[CRE 96] CRESSIE N.A., "Change of support and the modifiable areal unit problem", *Geographical Systems*, vol. 3, nos 2–3, pp. 159–180, 1996.

[CUZ 90] CUZICK J., EDWARDS R., "Spatial clustering for inhomogeneous populations", *Journal of the Royal Statistical Society, Series B*, vol. 52, pp. 73–104, 1990.

[DEA 92] DEAN C.B., "Testing for overdispersion in Poisson and Binomial regression models", *JASA*, vol. 87, no. 418, pp. 451–457, 1992.

[DIE 00] DIEKMANN O., HEESTERBEEK J.A.P., *Mathematical Epidemiology of Infectious Diseases: Model Building, Analysis and Interpretation*, Wiley, 2000.

[DIG 83] DIGGLE P.J., *Statistical Analysis of Spatial Point Patterns*, Academic Press, London, 1983.

[DIG 98] DIGGLE P.J., TAWN J.A., MOYEED RA., "Model-based geostatistics", *Applied Statistics*, vol. 47, Part 3, pp. 299–350, 1998.

[DIG 03] DIGGLE P.J., *Statistical Analysis of Spatial Point Patterns*, Hodder Arnold Publishers, London, 2003.

[DIX 94] DIXON P.M., "Testing spatial segregation using a nearest neighbor contingency table", *Ecology*, vol. 75, pp. 1940–1948, 1994.

[DIX 02] DIXON P.M., "Nearest neighbour methods", *Encyclopedia of Environmetrics*, vol. 3, John Wiley & Sons, pp. 1370–1383, 2002.

[DOR 07] DORMANN C., MCPHERSON J., ARAÚJO M. *et al.*, "Methods to account for spatial autocorrelation in the analysis of species distributional data: a review", *Ecography*, vol. 30, pp. 609–628, 2007.

[DRO 06] DROESBEKE J.J., LEJEUNE M., SAPORTA M., *Analyse statistique des données spatiales*, Technip Editions, 2006.

[ELI 08] ELITH J., LEATHWICK J.R., HASTIE T., "A working guide to boosted regression trees", *Journal of Animal Ecology*, vol. 77, pp. 802–813, 2008.

[ELL 92] ELLIOT P., CUZICK J., ENGLISH D. *et al.*, *Geographical & Environmental Epidemiology, Methods for Small-area Studies*, Oxford, 1992.

[ELL 00] ELLIOT P., WAKEFIELD J., BEST N. *et al.*, *Spatial Epidemiology, Methods and Applications*, Oxford, 2000.

[FOT 02] FOTHERINGHAM A.S., BRUNSDON C., CHARLTON M., *Geographically Weighted Regression: The Analysis of Spatially Varying Relationships*, Wiley, 2002.

[FOT 09] FOTHERINGHAM A.S., ROGERSON PA (eds), *The Sage Handbook of Spatial Analysis*, Sage, London and Los Angeles, 2009.

[FOT 15] FOTHERINGHAM A.S., CRESPO R., YAO J., "Geographical and temporal weighted regression (GTWR)", *Geographical Analysis*, vol. 47, pp. 431–452, 2015.

[GAL 04] LE GALLO J., "Hétérogénéité spatiale : principes et méthodes", *Économie & Prévision*, no. 162, pp. 151–172, 2004.

[GAT 96] GATRELL A.C., BAILEY T.C., DIGGLE P.J. *et al.*, "Spatial point pattern analysis and its application in geographical epidemiology", *Transactions of the Institute of British Geography*, vol. 21, pp. 256–274, 1996.

[GEA 54] GEARY R.C., "The contiguity ratio and statistical mapping", *The Incorporated Statistician*, vol. 5, pp. 115–145, 1954.

[GEL 10] GELFAND A., DIGGLE P., FUENTES M. *et al.*, *Handbook of Spatial Statistics*, Chapman & Hall/CRC, 2010.

[GET 96] GETIS A., ORD J.K., "Local spatial statistics: an overview", in LONGLEY P., BATTY M. (eds), *Spatial Analysis: Modelling in a GIS Environment*, Geoinformation International, Cambridge, pp. 261–277, 1996.

[GIN 21] GINI C., "Measurement of inequality of income", *Economic Journal*, vol. 31, pp. 22–43, 1921.

[GOT 02] GOTWAY C., YOUNG L., "Combining incompatible spatial data", *Journal of the American Statistical Association*, vol. 97, pp. 632–648, 2002.

[GRA 94] GRASLAND C., "L'erreur écologique en géographie", *Feuilles de Géographie*, vol. I, no. 1, 1994.

[HAR 10] HARDISTY F., KLIPPER A., "Analyzing spatio-temporal autocorrelation with LISTA-Viz", *International Journal of Geographic Information Sciences*, vol. 24, no. 10, pp. 1515–1526, 2010.

[HAS 90] HASTIE T.J., TIBSHIRANI R., *Generalized Additive Models*, Chapman and Hall, London, 1990.

[HAS 15] HASSARANGSEE S., TRIPATHI N., SOURIS M., Spatial Pattern Detection of Tuberculosis: A Case Study of Si Sa Ket Province, Thailand, IJERPH, 12 16005-16018, 2015.

[HAS 18] HASSARANGSEE S., Dissertation for the Degree of Doctor of Technical Science, Asian Institute of Technology, Thailand, 2018.

[HER 07] HERBRETEAU V., SALEM G., SOURIS M. *et al.*, "Thirty years of use and improvement of remote sensing, applied to epidemiology: from early promise to lasting frustration", *Health and Place*, vol. 13, pp. 400–403, 2007.

[HUA 07] HUANG L., KULLDORFF M., GREGORIO D., *A Spatial Scan Statistic for Survival data*, Biometrics, Wiley Online Library, 2007.

[ISA 60] ISARD W., *Methods of Regional Analysis: An Introduction to Regional Science*, The MIT Press, Cambridge, 1960.

[JUN 06] JUNG I., KULLDORFF K., KLASSEN A.C., "A spatial scan statistic for ordinal data", *Statistics in Medicine*, vol. 26, pp. 1594–1607, 2006.

[JUN 10] JUNG I., KULLDORFF M., OTUKEI J., "A spatial scan statistic for multinomial data", *Statistics in Medicine*, vol. 29, pp. 1910–1918, 2010.

[KLE 94] KLEINBAUM D.G., KLEIN M., *Statistics for Biology and Health*, Springer, 1994.

[KNE 05] KNEIB T., Mixed model based inference in structured additive regression model, Dissertation an der Fakultät für Mathematik, Informatik und Statistik der Ludwig-Maximilians-Universitä, Munich, 2005.

[KNO 64] KNOX G., "The detection of space-time interactions", *Applied Statistics*, vol. 13, pp. 25–29, 1964.

[KUL 95] KULLDORFF M., NAGARWALLA N., "Spatial disease clusters: detection and inference", *Statistics in Medicine*, vol. 14, no. 8, pp. 799–819, 1995.

[KUL 97] KULLDORFF M., "A spatial scan statistic", *Communications in Statistics: Theory and Methods*, vol. 26, no. 6, pp. 1481–1496, 1997.

[KUL 06] KULLDORF M., "Tests of spatial randomness adjusted for an inhomogeneity", *Journal of the American Statistical* Association, vol. 101, no. 475, pp. 1289–1305, 2006.

[KUL 09] KULLDORF M., HAUNG L., KONTY K., "A scan statistic for continuous data based on the normal probability model", *International Journal of Health Geography*, vol. 8, 2009.

[LAN 14] LANG S., UMLAUF N., WECHSELBERGER P. *et al.*, "Multilevel structured additive regression", *Statistics and Computing*, vol. 24, 2014.

[LAR 11] LARMARANGE J., VALLO R., YARO S. *et al.*, Méthodes pour cartographier les tendances régionales de la prévalence du VIH à partir des Enquêtes Démographiques et de Santé (EDS), *Cybergeo: European Journal of Geography*, http://cybergeo.revues.org/237 82, 2011.

[LAW 99] LAWSON A.B., BIGGERI A., BÖHNING D. *et al.* (eds), *Disease Mapping and Risk Assessment for Public Health*, John Wiley & Sons Ltd., 1999.

[LAW 01] LAWSON A.B., *Statisticial Methods in Spatial Epidemiology*, Wiley, 2001.

[LEE 01] LEE S.I., "Developing a bivariate spatial association measure: an integration of Pearson's r and Moran's I", *Journal of Geographical Systems*, vol. 3, no. 4, pp. 369–385, 2001.

[LEP 94] DE LEPPER J.C., SCHOLTEN H.J., STERN R.M., "The added value of geographical Information systems in public and environemental health", *The Geojournal Library*, vol. 24, 1994.

[LIE 96] VAN LIESHOUT M.N.M., BADDELEY A.J., "A nonparametric measure of spatial interaction in point patterns", *Statistica Neerlandica*, vol. 50, pp. 344–361, 1996.

[MAG 05] MAGUIRE D.J., BATTY M., GOODCHILD M., *GIS, Spatial Analysis and Modeling*, ESRI Press, 2005.

[MAN 67] MANTEL N., "The detection of disease clustering and a generalized regression approach", *Cancer Research*, vol. 27, no. 2, pp. 209–220, 1967.

[MAN 86] MANLY B.F.J., *Multivariate Statistical Methods, A Primer*, Chapman & Hall, 1986.

[MAN 16] MANEERAT S., Modélisation à base d'agents des risques vectoriels en milieux urbains : exemple d'Aedes aegypti, vecteur de la dengue, à Delhi (Inde), PhD Thesis, Université de Rouen, 2016.

[MAS 03] MASSÉ R., SAINT-ARNAUD J., *Ethique et santé publique*, Presses Université de Laval, 2003.

[MCK 04] McKenzie F.E., Samba E.M., "The role of mathematical modeling in evidence-based malaria control", *The American Journal of Tropical Medicine and Hygiene*, vol. 71, pp. 94–96, 2004.

[MIL 96] Miller R.G., *Simultaneous Statistical Inference*, Springer Verlag, 1996.

[MOL 99] Mollié A., "Bayesian and Empirical Bayes approaches to disease mapping", in Lawson A., Biggeri A., Bohning D. (eds), *Disease Mapping and Risk Assessment*, Wiley, pp. 15–29, 1999.

[MON 93] Monmonier M.S., *Comment faire mentir les cartes*, Flammarion, 1993.

[MOR 48] Moran P., "The interpretation of statistical maps", *Journal of the Royal Statistical Society, Series B*, vol. 10, pp. 243–251, 1948.

[MRK 15] Mrkvička T., Myllymäki M., Hahn U., "Multiple monte carlo testing with applications in spatial point processes", available at: http://arxiv.org/abs/1506.01646, 2015.

[NAK 15] Nakaya T., "Semiparametric geographically weighted generalised linear modelling: the concept and implementation using GWR4", in Brunsdon C., Singleton A. (eds), *Geocomputation: A Practical Primer*, Sage Publication, pp. 201–220, 2015.

[OMS 02] OMS, Rapport sur la santé dans le monde 2002, chapitre 2, Organisation Mondiale de la Santé, 2002.

[OPE 79] Openshaw S., Taylor P., "A million or so correlation coefficients: three experiments on the modifiable areal unit problem", in Wrigley N. (ed.), *Statistical Applications in the Spatial Sciences*, Pion, London, pp. 127–144, 1979.

[OPE 87] Openshaw S., Charlton M., Wymer C. *et al.*, "A mark I geographical analysis machine for the automated analysis of point data sets", *International Journal of Geographical Information Systems*, vol. 1, pp. 335–358, 1987.

[PEN 92] Penttinen A., Stoyan D., Henttonen H.M., "Marked point processes in forest statistics", *Forest Science*, vol. 38, pp. 806–824, 1992.

[PER 06] Perry G., Miller B., Enright N., "A comparison of methods for the statistical analysis of spatial point patterns in plant ecology", *Plant Ecology*, vol. 187, pp. 59–82, 2006.

[PFE 08] Pfeiffer D., Robinson T., Stevenson M. *et al.*, *Spatial Analysis in Epidemiology*, Oxford University Press, 2008.

[PIE 61] Pielou E.C., "Segregation and symmetry in two-species populations as studied by nearestneighbor relationships", *Journal of Ecology*, vol. 49, pp. 255–269, 1961.

[POT 96a] Potthoff R.F., Whittinghill M., "Testing for homogeneity: I. The Binomial and Multinomial distributions", *Biometrika*, vol. 53, pp. 167–182, 1996.

[POT 96b] Potthoff R.F., Whittinghill M., "Testing for homogeinity: II. The Poisson distribution", *Biometrika*, vol. 53, pp. 183–190, 1996.

[PUM 97] PUMAIN D., SAINT-JULIEN T., *L'analyse spatiale, localisations dans l'espace*, Armand Colin, 1997.

[REI 13] REINER R.C., PERKINS T.A., BARKER C.M. *et al.*, "A systematic review of mathematical models of mosquito-borne pathogen transmission: 1970–2010", *Journal of The Royal Society Interface*, vol. 10, 20120921, 2013.

[RIC 98] RICAN S., "Cartographie des données épidémiologiques", *Cahier Santé*, vol. 8, pp. 461–470, 1998.

[RIP 81] RIPLEY B.D., *Spatial Statistics*, John Wiley & Sons, 1981.

[ROG 97] ROGERSON P.A., "Surveillance systems for monitoring the development of spatial patterns", *Statistics in Medicine*, vol. 16, pp. 2081–2093, 1997.

[SAA 08] SAATY T.L., "Decision making with the analytic hierarchy process", *International Journal of Services Sciences*, vol. 1, no. 1, 2008.

[SAM 91] SAMUELS S.J., BEAUMONT J.J., BRESLOW N.E., "Power and detectable risk of seven tests for SMR", *American Journal of Epidemiology*, vol. 133, pp. 1191–1197, 1991.

[SAN 07] SANDERS L., *Models in Spatial Analysis*, ISTE, London, 2007.

[SAT 18] SaTScan User Guide, SaTScan software manual, available at: www.satscan.org, 2018.

[SAV 17] SavGIS., Manuels et documentation, available at: www.savgis.org, 2017.

[SHA 05] SHABENBERGER O., GOTWAY C., *Statistical Methods for Spatial Data Analysis*, Chapman & Hall, 2005.

[SHE 01] SHEKHAR S., HUANG Y., "Discovering Spatial Co-location Patterns: A summary of Results", *Proceedings of 7th International Symposium on Spatial and Temporal Databases*, 2001.

[SNO 54] SNOW J., *On the Mode of Communication of Cholera*, Churchill Livingstone, 1854.

[SOU 02] SOURIS M., La construction d'un système d'information géographique, principes et algorithms, PhD Thesis, Université de La Rochelle, 2002.

[SOU 06] SOURIS M., "Les contraintes d'intégrité spatiale", in SERVIGNE S., LIBOUREL T. (eds), *Fondements des bases de données spatiales*, Hermès, pp. 99–123, 2006.

[SOU 10a] SOURIS M., GONZALEZ J.P., "Dynamique spatiale et temporelle de l'émergence du virus de l'influenza aviaire hautement pathogène. L'exemple des vagues épizootiques en Thaïlande", *Les maladies émergentes chez le végétal, l'animal et l'homme. Enjeux scientifiques et sociétaux, stratégies de détection et d'analyse*, Editions QUAE, 2010.

[SOU 10b] SOURIS M., GONZALEZ J.P., SHANMUGASUNDARAM J. *et al.*, "Retrospective space-time analysis of H5N1 Avian Influenza emergence in Thailand", *International Journal of Health Geographics*, vol. 9, no. 3, 2010.

[SOU 11] SOURIS M., BICHAUD L., "Statistical methods for bivariate spatial analysis in marked points. Examples in spatial epidemiology", *Journal of Spatial and Spatio-temporal Epidemiology*, vol. 2, no. 4, pp. 227–234, 2011.

[SOU 14] SOURIS M., SELENIC D., KHAKLANG S. *et al.*, "Poultry farms vulnerability and risk of Avian Influenza re-emergence in Thailand", *International Journal of Environmental Research and Public Health*, vol. 11, no. 1, pp. 934–951, 2014.

[SOU 17] SOURIS M., MARCOMBE S., LAFORET J. *et al.*, "Modeling spatial variation in risk of presence and insecticide resistance for malaria vectors in Laos", *PLoS ONE*, vol. 12, no. 5, e0177274 p., 2017.

[STO 88] STONE R.A., "Investigating of excess environmental risks around putative sources: Statistical problems and a proposed test", *Statistics in Medicine*, vol. 7, pp. 649–660, 1988.

[STO 95] STOYAN D., KENDALL W.S., MEKCE J., *Stochastic Geometry and Its Applications*, John Wiley & Sons, Chichester, 1995.

[STO 00] STOYAN D., PENTTINEN A., "Recent applications of point process methods in forestry statistics", *Statistical Science*, vol. 15, no. 1, pp. 61–73, 2000.

[TAN 95] TANGO T., "A class of tests for detecting general and focused clustering of rare diseases", *Statistics in Medicine*, vol. 14, pp. 2323–2334, 1995.

[TAN 02] TANGO T., "Score tests for detecting excess risks around putative sources", *Statistics in Medicine*, vol. 21, pp. 497–514, 2002.

[THE 71] THEIL H., FINEZZA A.J., "A note on the measurement of racial integration of schools by means of informational concepts", *Journal of Mathematical Sociology*, vol. 1, no. 2, pp. 187–194, 1971.

[TOB 70] TOBLER W., "A computer movie simulating urban growth in the Detroit region", *Economic Geography Supplement*, vol. 46, no. 2, pp. 234–240, 1970.

[TRZ 14] TRZASKA S., SCHNARR E., *A Review of Downscaling Methods for Climate Change Projections*, USAID, 2014.

[VIG 97] VIGNERON E., *Statistique et géographie*, PUF, 1997.

[WAL 95] WALLER L., LAWSON A., "The power of focused tests to detect disease clustering", *Statistics in Medicine*, vol. 14, pp. 2291–2308, 1995.

[WAL 96] WALLER L., "Statistical power and design of focused clustering studies", *Statistics in Medicine*, vol. 15, pp. 765–782, 1996.

[WAL 04] WALLER L., GOTWAY C.A., *Applied Spatial Statistics for Public Health Data*, Wiley, Hoboken, 2004.

[WAR 85] WARTENBERG D., "Multivariate spatial correlation: a method for exploratory geographical analysis", *Geographical Analysis*, vol. 17, pp. 263–283, 1985.

[WEB 90] WEBSTER R., OLIVER M.A., *Statistical Methods in Soil and Land Resource Survey*, Oxford University Press, Oxford, 1990.

[WOR 95] WORBOYS M.F., *GIS, a Computer Perspective*, Taylor and Francis, 1995.

[WU 14] WU B., LI R., HUANG B., "A geographically and temporally weighted autoregressive model with application to housing prices", *International Journal of Geographical Information Science*, vol. 28, pp. 1186–1204, 2014.

[XIE 08] XIE Z., TAN J., "Kernel density estimation of traffic accidents in a network space", *Computers, Environment and Urban Systems*, vol. 32, no. 5, pp. 396–406, 2008.

[YAM 07] YAMADA I., THILL J.C., "Local indicators of network-constrained clusters in spatial point patterns", *Geographical Analysis*, vol. 39, pp. 268–292, 2007.

[YU 10] YU D., "Exploring spatiotemporally varying regressed relationships: the geographically weighted panel regression analysis", *The International Archives of the Photogrammetry, Remote Sensing and Spatial Information Sciences*, vol. 38, Part II, pp. 134–139, 2010.

[YU 14] YU D., "Understanding regional development mechanisms in greater Beijing area, China, 1995–2001, from a Spatial–Temporal Perspective", *GeoJournal*, vol. 79, pp. 195–207, 2014.

Index

A, B, C

D, E, G

H, I, K

L, M, N

O, P, R

S, T

U, V

Other titles from

ISTE

in

Information Systems, Web and Pervasive Computing

2019

AUGEY Dominique, with the collaboration of ALCARAZ Marina
Digital Information Ecosystems: Smart Press

CHAMOUX Jean-Pierre
The Digital Era 2: Political Economy Revisited

GAUCHEREL Cédric, GOUYON Pierre-Henri, DESSALLES Jean-Louis
Information, The Hidden Side of Life

2018

ARDUIN Pierre-Emmanuel
Insider Threats
(Advances in Information Systems Set – Volume 10)

CARMÈS Maryse
Digital Organizations Manufacturing: Scripts, Performativity and Semiopolitics
(Intellectual Technologies Set – Volume 5)

CARRÉ Dominique, VIDAL Geneviève
Hyperconnectivity: Economical, Social and Environmental Challenges
(Computing and Connected Society Set – Volume 3)

CHAMOUX Jean-Pierre
The Digital Era 1: Big Data Stakes

DOUAY Nicolas
Urban Planning in the Digital Age
(Intellectual Technologies Set – Volume 6)

FABRE Renaud, BENSOUSSAN Alain
The Digital Factory for Knowledge: Production and Validation of Scientific Results

GAUDIN Thierry, LACROIX Dominique, MAUREL Marie-Christine, POMEROL Jean-Charles
Life Sciences, Information Sciences

GAYARD Laurent
Darknet: Geopolitics and Uses
(Computing and Connected Society Set – Volume 2)

IAFRATE Fernando
Artificial Intelligence and Big Data: The Birth of a New Intelligence
(Advances in Information Systems Set – Volume 8)

LE DEUFF Olivier
Digital Humanities: History and Development
(Intellectual Technologies Set – Volume 4)

MANDRAN Nadine
Traceable Human Experiment Design Research: Theoretical Model and Practical Guide
(Advances in Information Systems Set – Volume 9)

PIVERT Olivier
NoSQL Data Models: Trends and Challenges

ROCHET Claude
Smart Cities: Reality or Fiction

SAUVAGNARGUES Sophie
Decision-making in Crisis Situations: Research and Innovation for Optimal Training

SEDKAOUI Soraya
Data Analytics and Big Data

SZONIECKY Samuel
Ecosystems Knowledge: Modeling and Analysis Method for Information and Communication
(Digital Tools and Uses Set – Volume 6)

2017

BOUHAÏ Nasreddine, SALEH Imad
Internet of Things: Evolutions and Innovations
(Digital Tools and Uses Set – Volume 4)

DUONG Véronique
Baidu SEO: Challenges and Intricacies of Marketing in China

LESAS Anne-Marie, MIRANDA Serge
The Art and Science of NFC Programming
(Intellectual Technologies Set – Volume 3)

LIEM André
Prospective Ergonomics
(Human-Machine Interaction Set – Volume 4)

MARSAULT Xavier
Eco-generative Design for Early Stages of Architecture
(Architecture and Computer Science Set – Volume 1)

REYES-GARCIA Everardo
The Image-Interface: Graphical Supports for Visual Information
(Digital Tools and Uses Set – Volume 3)

REYES-GARCIA Everardo, BOUHAÏ Nasreddine
Designing Interactive Hypermedia Systems
(Digital Tools and Uses Set – Volume 2)

SAÏD Karim, BAHRI KORBI Fadia
Asymmetric Alliances and Information Systems:Issues and Prospects
(Advances in Information Systems Set – Volume 7)

SZONIECKY Samuel, BOUHAÏ Nasreddine
Collective Intelligence and Digital Archives: Towards Knowledge Ecosystems
(Digital Tools and Uses Set – Volume 1)

2016

BEN CHOUIKHA Mona
Organizational Design for Knowledge Management

BERTOLO David
Interactions on Digital Tablets in the Context of 3D Geometry Learning
(Human-Machine Interaction Set – Volume 2)

BOUVARD Patricia, SUZANNE Hervé
Collective Intelligence Development in Business

EL FALLAH SEGHROUCHNI Amal, ISHIKAWA Fuyuki, HÉRAULT Laurent, TOKUDA Hideyuki
Enablers for Smart Cities

FABRE Renaud, in collaboration with MESSERSCHMIDT-MARIET Quentin, HOLVOET Margot
New Challenges for Knowledge

GAUDIELLO Ilaria, ZIBETTI Elisabetta
Learning Robotics, with Robotics, by Robotics
(Human-Machine Interaction Set – Volume 3)

HENROTIN Joseph
The Art of War in the Network Age
(Intellectual Technologies Set – Volume 1)

KITAJIMA Munéo
Memory and Action Selection in Human–Machine Interaction
(Human–Machine Interaction Set – Volume 1)

LAGRAÑA Fernando
E-mail and Behavioral Changes: Uses and Misuses of Electronic Communications

LEIGNEL Jean-Louis, UNGARO Thierry, STAAR Adrien
Digital Transformation
(Advances in Information Systems Set – Volume 6)

NOYER Jean-Max
Transformation of Collective Intelligences
(Intellectual Technologies Set – Volume 2)

VENTRE Daniel
Information Warfare – 2nd edition

VITALIS André
The Uncertain Digital Revolution
(Computing and Connected Society Set – Volume 1)

2015

ARDUIN Pierre-Emmanuel, GRUNDSTEIN Michel, ROSENTHAL-SABROUX Camille
Information and Knowledge System
(Advances in Information Systems Set – Volume 2)

BÉRANGER Jérôme
Medical Information Systems Ethics

BRONNER Gérald
Belief and Misbelief Asymmetry on the Internet

IAFRATE Fernando
From Big Data to Smart Data
(Advances in Information Systems Set – Volume 1)

KRICHEN Saoussen, BEN JOUIDA Sihem
Supply Chain Management and its Applications in Computer Science

NEGRE Elsa
Information and Recommender Systems
(Advances in Information Systems Set – Volume 4)

POMEROL Jean-Charles, EPELBOIN Yves, THOURY Claire
MOOCs

SALLES Maryse
Decision-Making and the Information System
(Advances in Information Systems Set – Volume 3)

SAMARA Tarek
ERP and Information Systems: Integration or Disintegration
(Advances in Information Systems Set – Volume 5)

2014

DINET Jérôme
Information Retrieval in Digital Environments

HÉNO Raphaële, CHANDELIER Laure
3D Modeling of Buildings: Outstanding Sites

KEMBELLEC Gérald, CHARTRON Ghislaine, SALEH Imad
Recommender Systems

MATHIAN Hélène, SANDERS Lena
Spatio-temporal Approaches: Geographic Objects and Change Process

PLANTIN Jean-Christophe
Participatory Mapping

VENTRE Daniel
Chinese Cybersecurity and Defense

2013

BERNIK Igor
Cybercrime and Cyberwarfare

CAPET Philippe, DELAVALLADE Thomas
Information Evaluation

LEBRATY Jean-Fabrice, LOBRE-LEBRATY Katia
Crowdsourcing: One Step Beyond

SALLABERRY Christian
Geographical Information Retrieval in Textual Corpora

2012

BUCHER Bénédicte, LE BER Florence
Innovative Software Development in GIS

GAUSSIER Eric, YVON François
Textual Information Access

STOCKINGER Peter
Audiovisual Archives: Digital Text and Discourse Analysis

VENTRE Daniel
Cyber Conflict

2011

BANOS Arnaud, THÉVENIN Thomas
Geographical Information and Urban Transport Systems

DAUPHINÉ André
Fractal Geography

LEMBERGER Pirmin, MOREL Mederic
Managing Complexity of Information Systems

STOCKINGER Peter
Introduction to Audiovisual Archives

STOCKINGER Peter
Digital Audiovisual Archives

VENTRE Daniel
Cyberwar and Information Warfare

2010

BONNET Pierre
Enterprise Data Governance

BRUNET Roger
Sustainable Geography

CARREGA Pierre
Geographical Information and Climatology

CAUVIN Colette, ESCOBAR Francisco, SERRADJ Aziz
Thematic Cartography – 3-volume series
Thematic Cartography and Transformations – Volume 1
Cartography and the Impact of the Quantitative Revolution – Volume 2
New Approaches in Thematic Cartography – Volume 3

LANGLOIS Patrice
Simulation of Complex Systems in GIS

MATHIS Philippe
Graphs and Networks – 2nd edition

THERIAULT Marius, DES ROSIERS François
Modeling Urban Dynamics

2009

BONNET Pierre, DETAVERNIER Jean-Michel, VAUQUIER Dominique
Sustainable IT Architecture: the Progressive Way of Overhauling Information Systems with SOA

PAPY Fabrice
Information Science

RIVARD François, ABOU HARB Georges, MERET Philippe
The Transverse Information System

ROCHE Stéphane, CARON Claude
Organizational Facets of GIS

2008

BRUGNOT Gérard
Spatial Management of Risks

FINKE Gerd
Operations Research and Networks

GUERMOND Yves
Modeling Process in Geography

KANEVSKI Michael
Advanced Mapping of Environmental Data

MANOUVRIER Bernard, LAURENT Ménard
Application Integration: EAI, B2B, BPM and SOA

PAPY Fabrice
Digital Libraries

2007

DOBESCH Hartwig, DUMOLARD Pierre, DYRAS Izabela
Spatial Interpolation for Climate Data

SANDERS Lena
Models in Spatial Analysis

2006

CLIQUET Gérard
Geomarketing

CORNIOU Jean-Pierre
Looking Back and Going Forward in IT

DEVILLERS Rodolphe, JEANSOULIN Robert
Fundamentals of Spatial Data Quality

Printed and bound by CPI Group (UK) Ltd, Croydon, CR0 4YY

15/09/2026

14962351-0002